# Visualization for Social Data Science

**Visualization for Social Data Science** provides end-to-end skills in visual data analysis. The book demonstrates how data graphics and modern statistics can be used in tandem to process, explore, model and communicate data-driven social science. It is packed with detailed data analysis examples, pushing you to do visual data analysis. As well as introducing, and demonstrating with code, a wide range of data visualizations for exploring patterns in data, Visualization for Social Data Science shows how models can be integrated with graphics to emphasise important structure and de-emphasise spurious structure and the role of data graphics in scientific communication -- in building trust and integrity. Many of the book's influences are from data journalism, as well as information visualization and cartography.

Each chapter introduces statistical and graphical ideas for analysis, underpinned by real social science datasets. Those ideas are then implemented via principled, step-by-step, workflows in the programming environment R. Key features include:

- Extensive real-world data sets and data analysis scenarios in Geography, Public Health, Transportation, Political Science;
- Code examples fully-integrated into main text, with code that builds in complexity and sophistication;
- Quarto template files for each chapter to support literate programming practices;
- Functional programming examples, using tidyverse, for generating empirical statistics (bootstrap resamples, permutation tests) and working programmatically over model outputs;
- Unusual but important programming tricks for generating sophisticated data graphics such as network visualizations, dot-density maps, OD maps, glyphmaps, icon arrays, hypothetical outcome plots and graphical line-ups plots. Every data graphic in the book is implemented via ggplot2.
- Chapters on uncertainty visualization and data storytelling that are uniquely accompanied with detailed, worked examples.

**Roger Beecham** is Associate Professor of Visual Data Science at University of Leeds School of Geography and Director of Research & Innovation at Leeds Institute for Data Analytics. He has published award-winning methodological work in data visualization, statistical practice and applied social science. He has taught visual data analysis for many years – to undergraduate and postgraduate students, experienced academics and data analysis professionals.

Chapman & Hall/CRC
Statistics in the Social and Behavioral Sciences Series

**Series Editors**
Jeff Gill, Steven Heeringa, Wim J. van der Linden, and Tom Snijders

**Recently Published Titles**

**Applied Regularization Methods for the Social Sciences**
Holmes Finch

**An Introduction to the Rasch Model with Examples in R**
Rudolf Debelak, Carolin Stobl, and Matthew D. Zeigenfuse

**Regression Analysis in R: A Comprehensive View for the Social Sciences**
Jocelyn H. Bolin

**Intensive Longitudinal Analysis of Human Processes**
Kathleen M. Gates, Sy-Min Chow, and Peter C. M. Molenaar

**Applied Regression Modeling: Bayesian and Frequentist Analysis of Categorical and Limited Response Variables with R and Stan**
Jun Xu

**The Psychometrics of Standard Setting: Connecting Policy and Test Scores**
Mark Reckase

**Crime Mapping and Spatial Data Analysis using R**
Juanjo Medina and Reka Solymosi

**Computational Aspects of Psychometric Methods: With R**
Patricia Martinková and Adéla Hladká

**Principles of Psychological Assessment**
With Applied Examples in R
Isaac T. Petersen

**Multilevel Modeling Using R, Third Edition**
W. Holmes Finch, Jocelyn E. Bolin, and Ken Kelley

**Polling, Prediction, and Testing, Second Edition**
Ole J. Forsberg

**Generalized Kernel Equating with Applications in R**
Marie Wiberg, Jorge Gonzalez and Alina A. von Davier

**Applied Survey Data Analysis, Third Edition**
Brady T. West, Steven G. Heeringa, and Patricia A. Berglund

**Visualization for Social Data Science**
Roger Beecham

**Introduction to Bayesian Data Analysis for Cognitive Science**
Bruno Nicenboim, Daniel J. Schad and Shravan Vasishth

For more information about this series, please visit: https://www.routledge.com/Chapman--HallCRC-Statistics-in-the-Social-and-Behavioral-Sciences/book-series/CHST-SOBESCI

# Visualization for Social Data Science

Roger Beecham

**CRC Press**
Taylor & Francis Group
Boca Raton London New York

CRC Press is an imprint of the
Taylor & Francis Group, an **informa** business

A CHAPMAN & HALL BOOK

Designed cover image: Roger Beecham

First edition published 2026
by CRC Press
2385 NW Executive Center Drive, Suite 320, Boca Raton FL 33431

and by CRC Press
4 Park Square, Milton Park, Abingdon, Oxon, OX14 4RN

*CRC Press is an imprint of Taylor & Francis Group, LLC*

ISBN: 978-1-032-27437-9 (hbk)
ISBN: 978-1-032-25971-0 (pbk)
ISBN: 978-1-003-29276-0 (ebk)

DOI: 10.1201/9781003292760

Typeset in Latin Modern font
by KnowledgeWorks Global Ltd.

Publisher's Note: This book has been prepared from camera-ready copy provided by the author.

# Table of contents

# *Preface*

Social scientists have at their disposal an expanding array of data measuring very many social behaviours. This is undoubtedly a positive. Previously unmeasurable aspects of human behaviour can now be explored in a large-scale empirical way, while already measured aspects of behaviour can be re-evaluated. Such data are nevertheless rarely generated for the sole purpose of social research, and this fact elevates visual approaches in importance due to visualization's emphasis on discovery. When encountering new data for the first time, data graphics help expose complex structure and multivariate relations, and in so doing advance an analysis in situations where the questions to be asked and techniques to be deployed may not be immediately obvious.

Visualization toolkits such as `ggplot2`, `vega-lite` and `Tableau` have been designed to ease the process of generating data graphics for analysis. There is a comprehensive set of texts and resources on visualization design theory, and several notable *how-to* primers on visualization practice. However, comparatively few existing resources demonstrate with real data and real social science scenarios *how* and *why* data graphics should be incorporated in a data analysis, and ultimately how they can be used to generate and claim knowledge.

This book aims to fill this space. It presents principled workflows, with code, for using data graphics and statistics in tandem. In doing so it equips readers with critical design and technical skills needed to analyse and communicate with a range of datasets in the social sciences.

The book emphasises application. Each chapter introduces concepts for analysis, with an accompanying technical implementation that uses real-world data on a range of Public Health, Transportation, Social and Electoral outcomes. The ambition is that by the end of each chapter, we have a more advanced knowledge and understanding of the phenomena under investigation.

## Structure, content and outcomes

Chapters of the book are divided into *Concepts* and *Techniques*. The *Concepts* sections cover key literature, ideas and approaches that can be leveraged to analyse the dataset introduced in the chapter. In the *Techniques* sections, code examples are provided for implementing those concepts and ideas. Each chapter starts with a list of *Knowledge* and *Skills* outcomes that map to the *Concepts* and *Techniques*. To support the technical elements, chapters have a corresponding computational notebook file. These files contain pre-prepared code chunks to be executed. In the early chapters we aim at brevity in the *Concepts* sections, offset by slightly more lengthy *Techniques* sections. As the book progresses the balance shifts somewhat, with more involved conceptual discussion and more specialised and abbreviated technical demonstrations.

Readers of the book will learn how to:

- Describe, process and combine social science datasets from a range of sources.
- Design statistical graphics that expose structure in social science data and that are underpinned by evidence-backed practice in information visualization and cartography.
- Use data science and visualization frameworks to produce data analysis code that is coherent and easily shareable.
- Apply modern statistical and graphical techniques for analysing, representing and communicating data and model outputs with integrity.

---

## Audience and assumed background

The book is for people analysing societal issues, broadly defined, including from within Geography, Public Health, Transportation and Political Science. It is aimed at postgraduate students and researchers, data journalists, analysts working in public sector and commercial organisations.

All technical examples are implemented using the R programming environment; so too *every* data graphic that appears in this book. Some prior knowledge of the R ecosystem is assumed, and as the chapters progress, more advanced concepts and coding procedures are introduced. While the book covers many of the fundamentals of R for working with social science datasets, our ultimate aim is to demonstrate through example how data graphics can and should be used in a data analysis. In this way it complements core resources that more fully cover, from zero-level prior knowledge, these how-to aspects: *R for Data Science* (Wickham and Grolemund 2017), *Tidymodelling with R* (Kuhn and Silge 2023) and *Geocomputation with R* (Lovelace, Nowosad, and Muenchow 2019).

## Omissions and additions

There are certain aspects of the book that might be surpising to those seasoned in reading data visualization textbooks. We do not cover interactivity in data graphics, and there is not a chapter dedicated to geospatial visualization, though numerous geospatial visualizations (maps) appear throughout to address particular analysis questions.

The reasons for this are principled as well as pragmatic. The R programming environment is not well-suited to highly flexible, interactive data graphics. Even if it were, we would question the need for interaction in many of the real-world data analysis scenarios covered in this book. The lack of a dedicated geovisualization chapter will hopefully become clear by the end of Chapter 3. It is useful to apply the same theory, heuristics and coding ideas to designing and evaluating maps as one would any other data graphic.

Space in the book is instead dedicated to introspecting into data graphics: the role of statistics and models for emphasising important structure and de-emphaising spurious structure, the differing purposes of data graphics at different analysis stages and the role of data graphics in building trust and integrity. Many of the book's influences are from data journalism, as well as information visualization and cartography.

## Acknowledgments

You will notice that the book is written in the first person, but with "we/our" rather than the singular pronoun "I/my". The reasons for this are partly stylistic. They also, hopefully, betray that the ideas and work presented in the book are not entirely my own. In particular "I" would like to thank Jo Wood and Jason Dykes, whose thinking on visualization design and practice runs throughout the book; and Robin Lovelace, who helped get things kick-started, whose technical knowledge is legion and whose critique and encouragement is always welcome. Thanks also to Lara Spieker from CRC Press and Taylor & Francis for helping move from an early plan to full production. And finally, as ever, to the reviewers for providing expert feedback on the book's structure and emphasis, and for the more general encouragement and positivity.

# 1

## *Introduction*

By the end of this chapter you should gain the following knowledge and practical skills.

---

**Knowledge outcomes**

- ☐ Appreciate the motivation for this book: why visualization, why R and why `ggplot2`.
- ☐ Recognise the characteristics of reproducible research and the role of RStudio Projects and computational notebooks (Quarto) for curating data analysis reports.

---

**Skills outcomes**

- ☐ Open R using the RStudio Integrated Developer Environment (IDE).
- ☐ Install and enable R packages and query package documentation.
- ☐ Create R Projects.
- ☐ Read-in external datasets as in-memory data frames.
- ☐ Render Quarto files.

---

## 1.1 Introduction

This chapter introduces the *what*, *why* and *how* of the book. After defining Data Science in a way that hopefully resists hyperbole, we demonstrate the importance of visual approaches in modern data analysis, especially social science analysis. We then introduce the key technologies and analysis frameworks for the book. In the technical component we consolidate any prior knowledge of the R ecosystem, demonstrate how to organise data science analyses as RStudio Projects and how to curate data analysis reports as computational notebooks via Quarto.

## 1.2 Concepts

### 1.2.1 *Why* visualization?

It is now taken for granted that new data, new technology and new ways
of doing science have transformed how we approach the world's problems.
Evidence for this can be seen in the response to the Covid-19 pandemic. Enter
'*Covid19 github*' into a search and you'll be confronted with hundreds of code
repositories demonstrating how data related to the pandemic can be collected,
processed and analysed. Data Science (hereafter data science) is a catch-all
term used to capture this shift.

The definition has been somewhat stretched over the years, but data science
has its origins in the work of John Tukey's *The Future of Data Analysis* (1962).
Drawing on this, and a survey of more recent work, Donoho (2017) identifies
six key facets that a data science discipline might encompass:

1.  data gathering, preparation and exploration;
2.  data representation and transformation;
3.  computing with data;
4.  data visualization and presentation;
5.  data modelling;
6.  and a more introspective "science about data science".

Each is covered to varying degrees within the book. Data visualization and
presentation of course gets a special status. Rather than a single and self-
contained facet of data science process – something that happens after data
gathering, preparation and exploration, but before modelling – the book
demonstrates how data visualization is intrinsic to, or at least should inform,
every facet of data science work: to capture complex, multivariate structure
(Chapters 3, 4, 5), provoke critical thinking around data transformation and
modelling (Chapters 4, 5 and 6) and communicate observed patterns with
integrity (Chapters 7 and 8).

This special status is further justified when considering the circumstances under
which Social Data Science (hereafter social data science) projects operate. New
datasets are typically repurposed for social science research for the first time,
contain complex structure and relations that cannot be easily modelled using
conventional statistics and, as a consequence, the types of questions asked and
techniques deployed to answer them – the research design – cannot be easily
specified in advance.

## Case study: using data visualization for urban mobility analysis

Let's develop this argument through example. In the early 2010s, several major cities around the world launched large-scale bikeshare systems. Data harvested from these systems enable city-wide cycling behaviours to be profiled in new ways, but they also present challenges. Bikeshare systems describe a particular category of cycling. The available user data, while spatiotemporally precise and 'population-level', are insufficiently detailed to easily assess how typical of cyclists are their users. Factors such as motivations, drivers and barriers to cycling, which especially interest transport researchers and planners, can only be inferred since they are not measured directly.

Figure 1.1 contains a sample of user data collected via London's bikeshare system. The Journeys table describes individual trips made between bikeshare docking stations; Stations, the locations of docking stations; and Members, high-level details of system users that can be linked to Journeys via a memberID. Figure 1.1 also shows statistical summaries that help us guess at how the system might be used: the hourly and daily profile of trips implying commuter-oriented usage; the 1D distribution of journey frequencies suggesting short, so-called 'last mile' trips; the expected heavy-tail in the rank-size plot confirming a large share of trips are made between a relatively select set of docking stations.

**Journeys**

| memberID | isMember | oTime | dTime | oStation | dStation |
|---|---|---|---|---|---|
| INTEGER | BOOLEAN | DATETIME | DATETIME | INTEGER | INTEGER |
| ##### | 1 | 30/07/2013 03:35 | 30/07/2013 03:41 | 308 | 308 |
| ##### | 1 | 30/07/2013 04:05 | 30/07/2013 04:21 | 290 | 286 |
| ##### | 1 | 30/07/2013 04:22 | 30/07/2013 04:42 | 81 | 174 |
| ##### | 1 | 30/07/2013 04:37 | 30/07/2013 04:47 | 14 | 14 |
| ##### | 1 | 30/07/2013 04:39 | 01/08/2013 04:10 | 169 | 169 |

**Stations**

| stationID | shortName | easting | northing | capacity |
|---|---|---|---|---|
| INTEGER | TEXT | INTEGER | INTEGER | INTEGER |
| 1 | River St\nEC2 | 531202 | 182838 | 19 |
| 2 | Phillimore Gdns\nW8 | 525207 | 179398 | 37 |
| 3 | Christopher St\nEC2 | 532984 | 182007 | 32 |
| 4 | St. Chad's St\nWC1 | 530436 | 182918 | 23 |
| 5 | Sedding St\nSW1 | 528050 | 178800 | 27 |

**Members**

| memberID | gender | regDate | postcode | easting | northing |
|---|---|---|---|---|---|
| INTEGER | TEXT | DATE | TEXT | INTEGER | INTEGER |
| ##### | F | 05/08/2011 | NW5### | 530## | 180### |
| ##### | M | 22/08/2012 | E10### | 540### | 190### |
| ##### | M | 28/01/2011 | N6### | 530### | 190### |
| ##### | M | 04/12/2012 | BN1### | 530### | 110### |
| ##### | F | 03/03/2011 | RH6### | 530### | 140### |

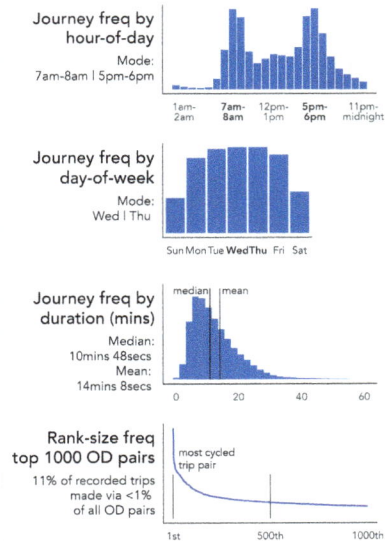

Journey freq by hour-of-day — Mode: 7am-8am | 5pm-6pm (1am-2am, 7am-8am, 12pm-1pm, 5pm-6pm, 11pm-midnight)

Journey freq by day-of-week — Mode: Wed | Thu (Sun Mon Tue Wed Thu Fri Sat)

Journey freq by duration (mins) — median | mean; Median: 10mins 48secs; Mean: 14mins 8secs (0 20 40 60)

Rank-size freq top 1000 OD pairs — most cycled trip pair; 11% of recorded trips made via <1% of all OD pairs (1st 500th 1000th)

**Figure 1.1:** Database schemas and summaries of London bikeshare user data. The values in these table excerpts are entirely synthetic.

While useful, the summaries and statistical graphics in Figure 1.1 are abstractions. They do not necessarily characterise how users of the bikeshare system cycle around the city. With the variables available to us, locations and

timestamps describing the start and end of bikeshare trips, we can create
graphics that expose these more synoptic patterns of usage. In Figure 1.2, journeys that occur during the morning weekday peak are encoded using flow-lines
that curve towards their destination. To emphasise the most frequently cycled
journeys, the thickness and transparency of flow-lines is adjusted according
to trip frequency. From this, we get a more direct sense of city-wide cycling
behaviour: a clear commuter function in the morning peak, with trips from
London's major rail hubs – King's Cross and Waterloo – connecting central
and City of London.

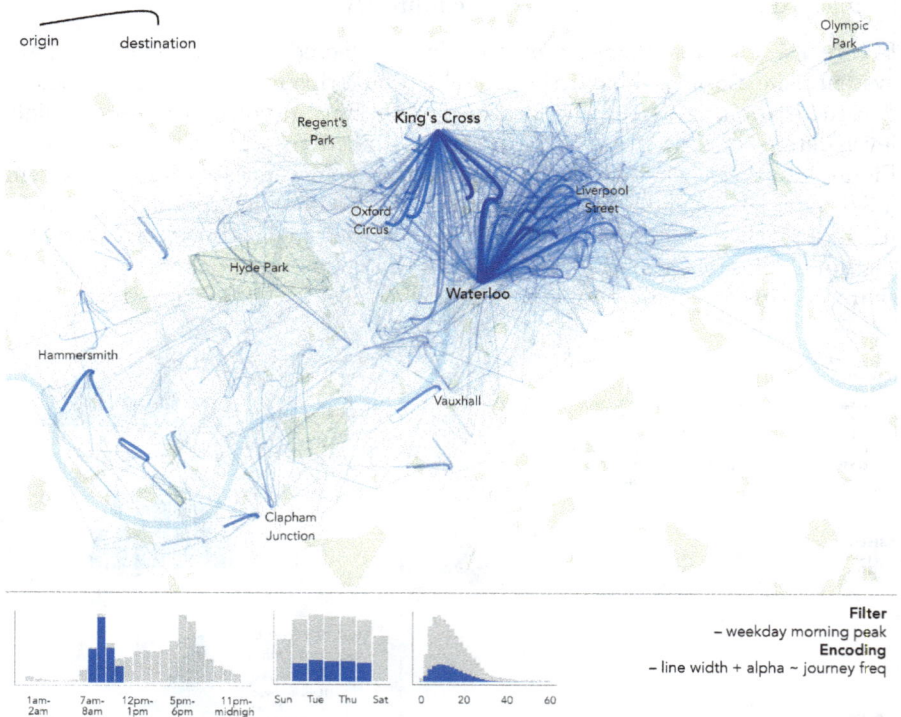

**Figure 1.2:** London bikeshare trips in 2018 (c. 10m records). Journeys are
filtered on the weekday morning peak.

The point of this example is not to undermine the value of statistical abstractions. Numeric summaries that simplify patterns are extremely useful and
Statistics has at its disposal an array of tools for helping to guard against
making false claims from datasets. There are, though, certain classes of relation
and context that especially pertain to social phenomena, geographic context
undoubtedly, that cannot be easily captured through numeric summary alone.

## 1.2.2   *What* type of visualization?

This book is as much about the role of visualization in statistics and data analysis practice as it is about the mechanics of data visualization. It leans heavily on real-world data and research questions. Each chapter starts with *concepts* for analysis, discussed via a specific dataset from the Political Science, Urban and Transport Planning and Health domains. Analysis of these data is then implemented via a *techniques* section. By the end of each chapter, we have a more advanced understanding of the phenomena under investigation, as well as an expanded awareness of visual data analysis practice.

To do this, we must cover a reasonably broad set of data processing and analysis procedures. As well as developing expertise on the design of data-rich, visually compelling graphics, some tedious aspects of data processing and wrangling are required. Additionally, to learn how to make and communicate claims under uncertainty with data graphics, techniques for estimation and modelling from Statistics are needed. In short, Donoho (2017)'s six key facets of a data science discipline:

1. data gathering, preparation, and exploration (Chapters 2, 3, 4);
2. data representation and transformation (Chapters 2, 3);
3. computing with data (Chapter 2, All chapters);
4. data visualization and presentation (All chapters);
5. data modelling (Chapters 4, 6, 7);
6. and the "science about data science" (All chapters).

### Case study: combining data graphics with models in urban mobility analysis

To demonstrate this more expanded role of visual data analysis, let's return to our bikeshare case study. Gender is an important theme in urban cycling research. High-cycling cities typically have equity in the level of cycling undertaken by men and women, and so the extent and nature of gender-imbalance can indicate how amenable to cycling a particular urban environment is. Of all trips made by members of London's bikeshare system, 77% are contributed by men. An obvious follow-up is whether the type and geography of these trips is distinctive.

If there were no differences in the trips made by men and women, we could set up a model that *expects* men to account for 77% of journeys cycled in any randomly sampled origin-destination (OD) journey pair in the dataset (the 'global' proportion of trips contributed by men).

In the rank-size plot below (Figure 1.3), we select out the top 50 most cycled OD pairs in the dataset and examine the male-female split against this expectation – the dark line. In only three of those ODs, in bold, do we see a higher than expected proportion of trips contributed by women. This suggests that the

journeys most popular with men are different from the journeys most popular
with women.

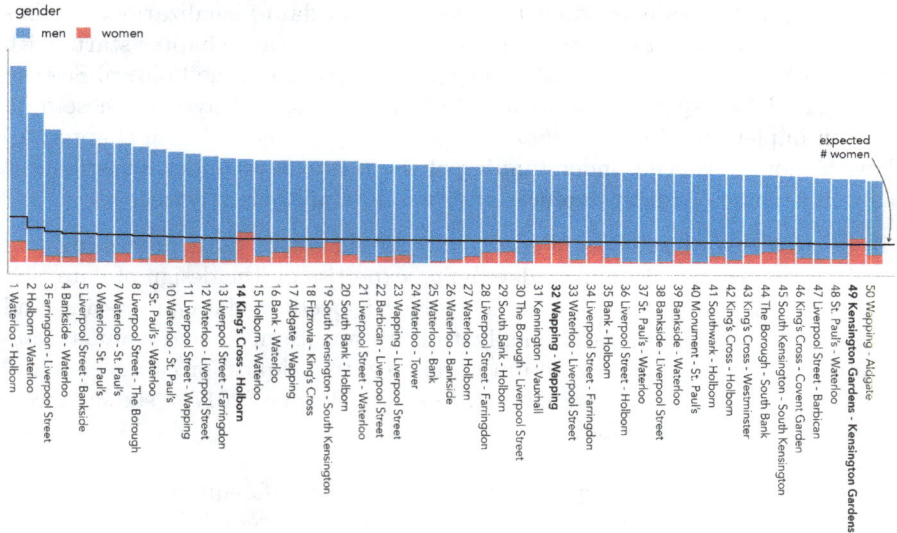

**Figure 1.3:** London bikeshare trips in 2018 (c. 10m records). Journeys are
filtered on the weekday morning peak.

To consider the spatial context behind these differences, and for a much larger
set of journeys, we update the flow-map graphic this time colouring flow lines
according the direction and extent of deviation from our modelled expectation
(Figure 1.4). The graphic shows stark geographic differences with men very
much overrepresented in bikeshare trips characteristic of commuting (dark
blue) – trips from major rail hubs (Waterloo and King's Cross) and city
and central London. By contrast women's travel behaviours are in fact more
geographically diverse and varied: the dark red emphasising OD pairs where
women are overrepresented.

The model in Figure 1.4 is not a particularly sophisticated one. A next step
would be to update it with important conditioning context that likely accounts
for some of the behavioural differences (see Beecham and Wood 2014). The act
of creating a model and encoding the original flow map with individual model
residuals, rather than relying on some global statistic of gender imbalance in
spatial cycling behaviour, is nevertheless clearly instructive. We will implement
this sort of visual data analysis throughout the book: 1. expose pattern; 2.
model an expectation derived from this pattern; 3. characterise deviation from
expectation.

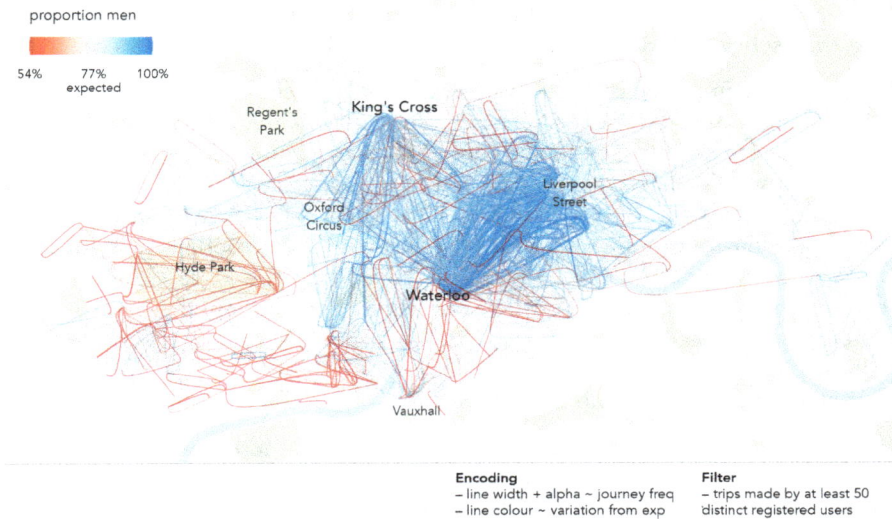

**Figure 1.4:** Gender comparison of London bikeshare trips made by registered users, Jan 2012 – Feb 2013 (c.5m records).

---

Task

Watch Jo Wood's TEDx talk, which gives a fuller explanation of this case study:

- `https://www.youtube.com/embed/FaRBUnO5PZI`

In the talk Jo argues that bikeshare systems can help democratise cycling, and he makes a compelling case for the role of visualization in uncovering structure in these sorts of large-scale behavioural data. You also might notice that Jo uses several rhetorical devices to communicate with data graphics; we will look deeper into these in Chapter 8, on Data Storytelling.

---

### 1.2.3 *How* we do visualization design and analysis

#### R for modern data analysis

All data collection, analysis and reporting activity will be completed using the open source statistical programming environment R. There are several benefits that come from being fully open-source, with a critical mass of users. Firstly, there is an array of online fora, tutorials and code examples from which to learn. Second, with such a large community, there are numerous expert R users who themselves contribute by developing packages that extend its use.

Of particular importance is the `tidyverse`. This is a set of packages for doing data science authored by the software development team at *Posit*. `tidyverse` packages share a principled underlying philosophy, syntax and documentation. Contained within the `tidyverse` is its data visualization package, `ggplot2`. This package predates the `tidyverse` and is one of the most widely-used toolkits for generating data graphics. As with other visualization toolkits it is inspired by Wilkinson (1999)'s *The Grammar of Graphics*; the `gg` in `ggplot2` stands for *Grammar of Graphics*. We will cover some of the design principles behind `ggplot2` and `tidyverse` in Chapter 3.

### Quarto for reproducible research

In the last decade there has been much introspection into how science works, particularly how statistical claims are made from reasoning over evidence. This came on the back of, amongst other things, a high profile paper published in the journal *Science* (Open Science Collaboration 2015), which found that of 100 contemporary peer-reviewed empirical papers in Psychology the findings of only 39 could be replicated. The upshot is that researchers must now endeavour to make their work transparent, such that "*all* aspects of the answer generated by any given analysis [can] be tested" (Brunsdon and Comber 2021).

A reproducible research project should be accompanied with *code* and *data* that:

- allow tables and figures presented in research outputs to be regenerated
- do what is claimed (the code works)
- can be justified and explained through proper documentation

In this setting proprietary data analysis software that support point-and-click interaction, previously used widely in the social sciences, are problematic. First, point-and-click software are usually underpinned by code that is closed. It is not possible, and therefore less common, for the researcher to fully interrogate the underlying procedures that are being implemented, and the results need to be taken more or less on faith. Second, replicating and updating the analysis in light of new data is challenging. It would be tedious to make notes describing all interactions performed when working with a dataset via a point-and-click-interface.

As a declarative programming environment, it is very easy to provide such a provenance trail in R. Also, and significantly, the Integrated Development Environments (IDEs) through which R is accessed offer computational notebook environments that blend *input code, explanatory prose* and *outputs*. Through the technical elements of this book, we will prepare these sorts of notebooks using Quarto.

## 1.3 Techniques

Readers of this book might already have some familiarity with R and the RStudio IDE. If not, then this section is designed to quickly acclimatise you to R and RStudio and to briefly introduce Quarto, R scripts and RStudio Projects. The accompanying template file, `01-template.qmd`[1], can be downloaded from the book's companion website. This material on setup and basics is introduced briskly. For a more involved introduction, readers should consult Wickham, Çetinkaya-Rundel, and Grolemund (2023), *the* handbook for data analysis in R.

### 1.3.1 R and RStudio

- Install the latest version of R. Note that there are installations for Windows, macOS and Linux. Run the installation from the file you downloaded (an `.exe` or `.pkg` extension).
- Install the latest version of RStudio Desktop. Note again that there are separate installations depending on operating system – for Windows an `.exe` extension, macOS a `.dmg` extension.
- Once installed, open the RStudio IDE.
- Open an R Script by clicking `File` > `New File` > `R Script` .

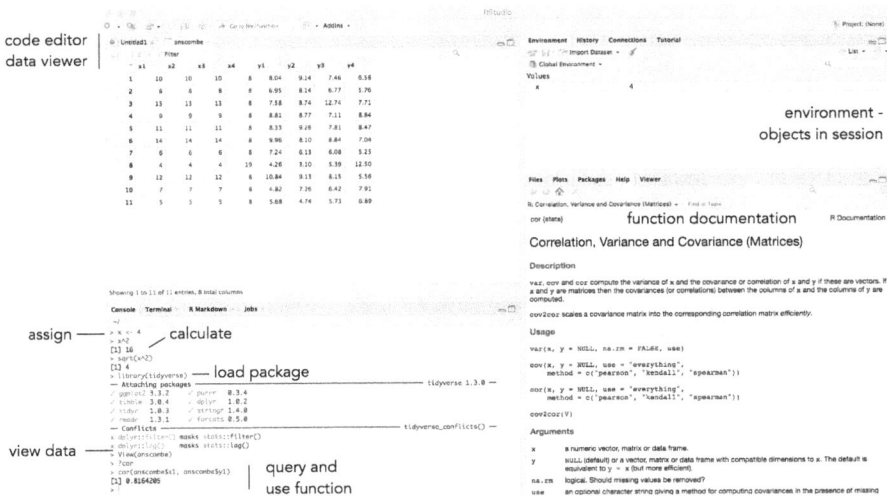

**Figure 1.5:** Annotated screenshots of the RStudio IDE.

You should see a set of windows roughly similar to those in Figure 1.5. The top left pane is used either as a code editor (the tab named `Untitled1`) or data

---

[1] https://vis4sds.github.io/vis4sds/files/01-template.qmd

viewer. This is where you'll write, organise and comment R code for execution or inspect datasets as a spreadsheet representation. Below this in the bottom left pane is the R *Console*, in which you write and execute commands directly. To the top right is a pane with the tabs *Environment* and *History*. This displays all objects – data and plot items, calculated functions – stored in-memory during an R session. In the bottom right is a pane for navigating through project folders, displaying plots, details of installed and loaded packages and documentation on functions.

### 1.3.2   Compute in the console

You will write and execute almost all code from the code editor pane. To start though, let's use R as a calculator by typing some commands into the console. You'll create an object (x) and assign it a value using the assignment operator (<-), then perform some simple statistical calculations using functions that are held within the base package.

Type the commands contained in the code block below into your R Console. Notice that since you are assigning values to each of these objects, they are stored in memory and appear under the Global Environment pane.

```
# Create variable and assign it a value.
x <- 4
# Perform some calculations using R as a calculator.
x_2 <- x^2
# Perform some calculations using functions that form base R.
x_root <- sqrt(x_2)
```

> **i** R package documentation
>
> The base package exists as standard in R. Unlike other packages, it does not need to be installed and called explicitly. One means of checking the package to which a function belongs is to call the documentation on that function, via the help command (?): e.g. ?mean().

### 1.3.3   Install some packages

There are two steps to getting packages down and available in your working environment:

1. `install.packages("<package-name>")` downloads the named package from a repository.
2. `library(<package-name>)` makes the package available in your current session.

Download `tidyverse`, the core collection of packages for doing Data Science in R, by running the code below:

```
install.packages("tidyverse")
```

If you have little or no experience in R, it is easy to get confused about downloading and then using packages in a session. For example, let's say we want to make use of the Simple Features package (`sf`) (Pebesma 2018) for performing spatial operations.

```
library(sf)
```

Unless you've previously installed `sf`, you'll probably get an error message that looks like this:

```
> Error in library(sf): there is no package called 'sf'
```

So let's install it.

```
install.packages("sf")
```

And now it's installed, bring up some documentation on one of its functions, `st_contains()`, by typing `?<function-name>` into the Console.

```
?st_contains()
```

Since you've downloaded the package but not made it available to your session, you should get the message:

```
> Error in .helpForCall(topicExpr, parent.frame()) :
  no methods for 'st_contains' and no documentation for it as a function
```

Let's try again, by first calling `library(sf)`. `?st_contains()` should execute without error, and the documentation on that function should be presented to you via the *Help* pane.

```
library(sf)
## Linking to GEOS 3.11.0, GDAL 3.5.3, PROJ 9.1.0
?st_contains()
```

So now install some of the remaining core packages on which this book depends. Run the block below, which passes a vector of package names to the `install.packages()` function.

```
pkgs <- c(
  "devtools","here", "quarto","fst","tidyverse", "lubridate",
  "tidymodels", "gganimate", "ggforce", "distributional", "ggdist"
)
install.packages(pkgs)
```

> **i R package visibility**
>
> If you wanted to make use of an installed package only very occasionally
> in a single session, you could access it without explicitly loading it via
> `library(<package-name>)`, using this syntax: `<package-name>::<function_name>`,
> e.g. `sf::st_contains()`.

### 1.3.4   Experiment with Quarto

Quarto documents are suffixed with the extension `.qmd`. They resemble Markdown, a lightweight language designed to minimise tedious markup tags
(`<header></header>`) when preparing HTML documents. The idea is that you
trade some flexibility in the formatting of your HTML for ease of writing.
Working with Quarto documents feels very similar to Markdown. Sections are
denoted hierarchically with hashes (`#`, `##`, `###`) and emphasis using "`*`" symbols
(`*emphasis* **added**` reads *emphasis* **added**). Different from standard Markdown, Quarto documents can also contain code chunks to be run when the
document is rendered. They support the creation of reproducible, dynamic and
interactive notebooks. Dynamic and reproducible because the outputs may
change when there are changes to the underlying data; interactive because
they can execute not just R code blocks, but also Jupyter Widgets, Shiny and
Observable JS. Each chapter of this book has an accompanying Quarto file.
In later chapters you will use these to author computational notebooks that
blend code, analysis prose and outputs.

Download the `01-template.qmd` file for this chapter and open it in RStudio by
clicking `File > Open File ... > <your-downloads>/01-template.qmd`. Note that there
are two tabs that you can switch between when working with `.qmd` files. *Source*
retains markdown syntax (e.g. `#|##|###` for headings); *Visual* renders these tags
and allows you to, for example, perform formatting and build tables through
point-and-click utilities.

A quick anatomy of `.qmd` files :

- YAML - positioned at the head of the document, and contains metadata
  determining amongst other things the author details and the output format
  when rendered.
- TEXT - incorporated throughout to expand upon your analysis.

- CODE chunks - containing discrete blocks that are run when the .qmd file is rendered.

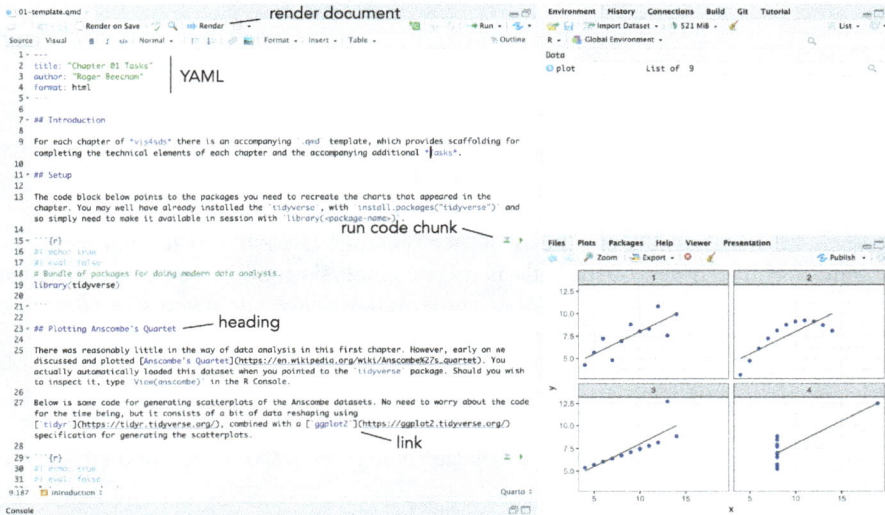

**Figure 1.6:** The anatomy of .qmd files

The YAML section of a .qmd file controls how your file is rendered and consists of key: value pairs enclosed by ---. Notice that you can change the output format to generate, for example, .pdf, .docx files for your reports.

```
---
author: "Roger Beecham"
title: "Chapter 01"
format: html
---
```

Quarto files are rendered with the *Render* button, annotated in Figure 1.6 above. This starts pandoc, a library that converts Markdown files, and executes all the code chunks and, in the case above, outputs an .html file.

- Render the 01-template.qmd file for this chapter by clicking the *Render* button.

Code chunks in Quarto can be customised in different ways. This is achieved by populating fields immediately after the curly brackets used to declare the code chunk.

```
```{r}
#| label: <chunk-name>
#| echo: true
#| eval: false
```
```

```
# The settings above mean that any R code below
# is not run (evaluated), but printed (echoed)
# in this position when the .qmd doc is rendered.
```
```

A quick overview of the parameters:

- `label: <chunk-name>` Chunks can be given distinct names. This is useful for navigating Quarto files. It also supports chaching – chunks with distinct names are only run once, important if certain chunks take some time to execute.
- `echo: <true|false>` Determines whether the code is visible or hidden from the rendered file. If the output file is a data analysis report, you may not wish to expose lengthy code chunks as these may disrupt the discursive text that appears outside of the code chunks.
- `eval: <true|false>` Determines whether the code is evaluated (executed). This is useful if you wish to present some code in your document for display purposes.
- `cache: <true|false>` Determines whether the results from the code chunk are cached.

### 1.3.5   R Scripts

While there are obvious benefits to working in `.qmd` documents for data analysis, there may be occasions where a script is preferable. R scripts are plain text files with the extension `.R`. They are typically used for writing discrete but substantial code blocks that are to be executed. For example, a set of functions that relate to a particular use case might be organised into an R script, and those functions referred to in a data analysis from a `.qmd` in a similar way as one might import a package. Below is an example script file with helper functions to support flow visualizations in R. The script is saved with the file name `bezier_path.R`. If it were stored in a sensible location, like a project's `code` folder, it could be called from a `.qmd` file with `source("code/bezier_path")`. R scripts can be edited in the same way as Quarto files in RStudio, via the code editor pane.

```
# Filename: bezier_path.R
# Author: Roger Beecham
#
#------------------------------------------------------------------------

# This function takes cartesian coordinates defining origin
# and destination locations and returns a tibble representing
# a path for an asymmetric bezier curve. The implementation
# follows Wood et al. 2011. doi: 10.3138/carto.46.4.239.
```
```

```r
# o_x, o_y : numeric coords of origin
# d_x, d_y : numeric coords of destination
# od_pair : text string identifying name of od-pair
# curve_extent : optional, controls curve angle
# curve_position : optional, controls curve position
get_trajectory <- function (o_x, o_y, d_x, d_y, od_pair,
  curve_extent=-90, curve_position=6)
  {
    curve_angle = get_radians(-curve_extent)
    x = (o_x - d_x)/curve_position
    y = (o_y - d_y)/curve_position
    c_x = d_x + x * cos(curve_angle) - y * sin(curve_angle)
    c_y = d_y + y * cos(curve_angle) + x * sin(curve_angle)
    d <- tibble::tibble(x = c(o_x, c_x, d_x), y = c(o_y, c_y,
        d_y), od_pair = od_pair)
    return(d)
  }


# Helper function converts degrees to radians.
# degrees : value of angle in degrees to be transformed
get_radians <- function(degrees) { (degrees * pi) / (180) }
```

R Scripts are more straightforward than Quarto files in that you don't have to worry about configuring code chunks. They are really useful for quickly developing bits of code. This can be achieved by highlighting the code that you wish to execute and clicking the Run icon at the top of the code editor pane or by typing `Ctrl` + `Rtn` on Windows, `Cmd` + `Rtn` on macOS.

> **i** .qmd, not R scripts, for data analysis
>
> Unlike .qmd, everything within a script is treated as code to be executed, unless it is commented with a #. Comments should be informative but paired back. As demonstrated above, it becomes somewhat tedious to read comments when they tend towards prose. For social science use cases, where code is largely written for analysis rather than software development, computational notebooks such as .qmd are preferred over R scripts.

## 1.3.6 Create an RStudio Project

Throughout the book we will use project-oriented workflows. This is where all files pertaining to a data analysis – data, code and outputs – are organised from a single top-level, or root, folder and where file path discipline is maintained such that all paths are relative to the project's root folder (see Chapter 7 of Wickham, Çetinkaya-Rundel, and Grolemund 2023). You can imagine this self-contained project setup is necessary for achieving reproducibility of your research. It allows anyone to take a project and run it on their own machines with minimal adjustment.

When opening RStudio, the IDE automatically points to a working directory, likely the home folder for your local machine. RStudio will save any outputs to this folder and expect any data you use to be saved there. Clearly, to incorporate neat, self-contained project workflows you will want a dedicated project folder rather than the default home folder for your machine. This can be achieved with the `setwd(<path-to-your-project>)` function. The problem with doing this is that you insert a path which cannot be understood outside of your local machine at the time it was created. This is a real pain. It makes simple things like moving projects around on your machine an arduous task, and most importantly it hinders reproducibility if others are to reuse your work.

RStudio Projects resolve these problems. Whenever you load an RStudio Project, R starts up and the working directory is automatically set to the project's root folder. If you were to move the project elsewhere on your machine, or to another machine, a new root is automatically generated – so RStudio projects ensure that relative paths work.

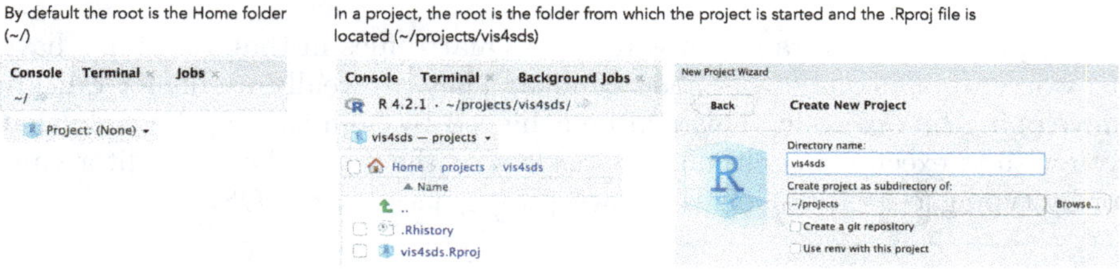

**Figure 1.7:** Creating an RStudio Project

Let's create a new Project for this book:

- Select `File` > `New Project` > `New Directory`.
- Browse to a sensible location and give the project a suitable name. Then click `Create Project`.

You will notice that the top of the Console window now indicates the root for this new project (`~projects/vis4sds`).

- In the top-level folder of your project, create folders called `code`, `data`, `figures`.
- Save this session's `01-template.qmd` file to the `vis4sds` folder.

Your project's folder structure should now look like this:

```
vis4sds\
  vis4sds.Rproj
  01-template.qmd
  code\
  data\
  figures\
```

## 1.4 Conclusions

Visual data analysis approaches are necessary for exploring complex patterns in data and to make and communicate claims under uncertainty. This is especially true of social data science applications, where datasets are repurposed for research often for the first time, contain complex structure and geo-spatial relations that cannot be easily captured by statistical summaries alone and, consequently, where the types of questions that can be asked and the techniques deployed to answer them cannot be specified in advance. This is demonstrated in the book as we explore (Chapters 4 and 5), model under uncertainty (Chapter 6) and communicate (Chapters 7 and 8) with various social science datasets. Different from other visualization 'primers', we pay particular attention to how *statistics* and *models* can be embedded into graphics (Gelman 2004). All technical activity in the book is completed in R, making use of tools and software libraries that form part of the R ecosystem: the tidyverse for doing modern data science and Quarto for helping to author reproducible research documents.

## 1.5 Further Reading

A paper that introduces modern data analysis and data science in a straight-forward way, eschewing much of the hype:

- Donoho, D. 2017. "50 Years of Data Science" *Journal of Computational and Graphical Statistics*, 26(6): 745–66. doi: 10.1080/10618600.2017.13847340.

An excellent 'live' handbook on reproducible data science:

- The Turing Way Community. 2022. The Turing Way: A handbook for reproducible, ethical and collaborative research (1.0.2). doi: 10.5281/zen-odo.3233853.

On R Projects and workflow:

- Wickham, H., Çetinkaya-Rundel, M., Grolemund, G. 2023, "R for Data Science, 2nd Edition", Sebastopol, CA: *O'Reilly*.
  - Chapter 6.

On Quarto:

- Wickham, H., Çetinkaya-Rundel, M., Grolemund, G. 2023, "R for Data Science, 2nd Edition", Sebastopol, CA: *O'Reilly*.
  - Chapters 28, 29.

# 2

## Data Fundamentals

By the end of this chapter you should gain the following knowledge and practical skills.

> **Knowledge outcomes**
>
> ☐ Learn the vocabulary and concepts used to describe data.
> ☐ Appreciate the characteristics and importance of tidy data (Wickham 2014).

> **Skills outcomes**
>
> ☐ Read-in large external files as data frames.
> ☐ Calculate descriptive summaries over datasets using `dplyr`.
> ☐ Learn how to structure, join and reshape data using `dplyr` and `tidyr`.
> ☐ Create statistical graphics for initial data description and exploration.

## 2.1 Introduction

This chapter covers the basics of how to describe and organise data. While this might sound prosaic, there are several reasons why being able to consistently describe a dataset is important. First, it is the initial step in any analysis and helps delimit the analytical procedures that can be deployed. This is especially relevant to modern data analysis, where it is common to apply the same analysis templates over many different datasets. Describing data using a consistent vocabulary enables you to identify which analysis templates to reuse. Second, relates to the point in Chapter 1, that social data science projects usually involve repurposing datasets for the first time. It is often not obvious whether a dataset contains sufficient detail and structure to characterise the behaviours being researched and the target populations it is assumed to represent. This leads to additional levels of uncertainty and places greater importance on the initial steps of data processing, description and exploration.

Through the chapter we will develop vocabulary for describing and thinking about data, as well as some of the most important data processing and organisation techniques in R. We will do so using data from New York's Citibike system.

> **i** Data vocabulary
>
> A consistent vocabulary for describing data is especially useful when learning modern visualization toolkits like `ggplot2`, `Tableau` and `vega-lite`. We will expand upon this in some detail in Chapter 3 as we introduce the fundamentals of visualization design and the Grammar of Graphics (Wilkinson 1999).

## 2.2 Concepts

### 2.2.1 Data frames

Throughout this book we will work with data frames. These are spreadsheet-like representations where rows are observations and columns are variables. In an R data frame, variables are vectors that must be of equal length. Where observations have missing values, for certain variables the missing values must be substituted with something, usually with NA or similar. This constraint can cause difficulties. For example, when working with variables that contain many values of different length for an observation, we create a special class of column, a `list-column`. Organising data according to this simple structure – rows as observations, columns as variables – is useful for developing analysis templates that work with the `tidyverse` package ecosystem.

### 2.2.2 Types of variable

A familiar classification for describing data is that developed by Stevens (1946) when considering the level of measurement of a variable. Stevens (1946) organises variables into two classes: variables that describe *categories* of things and variables that describe *measurements* of things.

Categories include attributes like gender, customer segments, ranked orders (1st, 2nd, 3rd largest etc.). Measurements include quantities like distance, age and travel time. Categories can be further subdivided into those that are unordered (*nominal*) and those that are ordered (*ordinal*). Measurements can also be subdivided. *Interval* measurements are quantities where the computed difference between two values is meaningful. *Ratio* measurements have this

**Table 2.1:** Breakdown of variable types and corresponding mathematical operations.

| Measurement | Example | Operators | Midpoint | Spread |
|---|---|---|---|---|
| Categories | | | | |
| Nominal | Political parties; street names | $= \neq$ | mode | entropy |
| Ordinal | Terrorism threat levels | $= \neq$ | median | percentile |
| Measures | | | | |
| Interval | Temperatures; years | $= \neq <> + -$ | mean | variance |
| Ratio | Distances; prices | $= \neq <> + -$ $\mid \times \div$ | mean | variance |

property, but also have a meaningful 0, where 0 means the absence of something, and the ratio of two values can be computed.

Why is this useful? The measurement level of a variable determines the types of data analysis operations that can be performed and therefore allows us to make quick decisions when working with a dataset for the first time (Table 2.1).

---

**Task 1**

Complete the data description table below identifying the *measurement level* of each variable in the New York bikeshare stations dataset.

| Variable name | Variable value | Measurement level |
|---|---|---|
| name | "Central Park" | |
| capacity | 80 | |
| rank_capacity | 45 | |
| date_opened | "2014-05-23" | |
| longitude | -74.00149746 | |
| latitude | 40.74177603 | |

---

## 2.2.3   Types of observation

Observations form an entire population or a sample that we expect is representative of a target population. In social data science applications we often work with datasets that are so-called population-level. The Citibike dataset is a complete, population-level dataset in that every Citibike journey is recorded. Whether or not this is truly a population-level dataset, however, depends on the analysis purpose. When analysing trips made by Citibike users, are we interested only in those cyclists? Or are we taking the patterns observed through our analysis to make inferences about New York cycling more generally? If the latter, then there are problems as the level of detail we have on our sample is pretty trivial compared to traditional, actively-collected datasets, where data collection activities are designed with a target population in mind. It may therefore be difficult to gauge how representative Citibike users and Citibike cycling is of New York's general cycling population. The flipside is that so-called passively-collected data may not suffer from the same problems of non-response bias and social-desirability bias as traditional, actively-collected data.

### 2.2.4   Tidy data

We will work with data frames organised such that columns always and only refer to variables and rows always and only refer to observations. This arrangement, called *tidy* (Wickham 2014), has two key advantages. First, if data are arranged in this tidy form, then it is easier to apply and re-use tools for wrangling them as they have the same underlying structure. Second, placing variables into columns with each column containing a vector of values, means that we can take advantage of R's vectorised functions for transforming data. This is demonstrated in the technical element of the chapter.

The three rules for tidy data (Wickham 2014):

1. Each variable forms a column.
2. Each observation forms a row.
3. Each type of observation unit forms a table.

The concept of tidy data, and its usefulness for `tidyverse`-style operations, is best explained through example. The technical element to this chapter is therefore comparatively lengthy and demonstrates key coding templates for organising and re-organising data frames for analysis.

---

## 2.3   Techniques

In this section we import, describe, transform and tidy data from New York's Citibike bikeshare system.

- Download the `02-template.qmd`[1] file, and save it to your `vis4sds` project, created in Chapter 1.
- Open your `vis4sds` project in RStudio, and load the template file by clicking `File > Open File ... > 02-template.qmd`.

---

[1]`https://vis4sds.github.io/vis4sds/files/02-template.qmd`

## 2.3.1 Import

In the template file there is documentation on how to set up your R session with key packages – `tidyverse` , `fst`, `lubridate`, `sf`. The data were collected using the `bikedata` R package. A subset of data from New York's bikeshare system, Citibike, were collected for this chapter and can be downloaded from the book's accompanying data repository[2].

The code for reading in these data may be familiar to most readers. The `here` package, which reliably creates paths relative to a project's root, is used to pass the locations at which the New York trips and stations data are stored as a parameter to `read_csv()` and `read_fst()`. Notice that we use *assignment* (<-), so data are loaded as objects and appear in the Environment pane of RStudio.

```
# Read in local copies of the trips and stations data.
ny_trips <- read_fst(here("data", "ny_trips.fst"))
ny_stations <- read_csv(here("data", "ny_stations.csv"))
```

> **i** `fst` for in-memory analysis
>
> `fst` is a special class of file that implements in the background various operations to speed-up reading and writing of data. This makes it possible to work with large datasets in-memory in R rather than connecting to a database and returning data summaries/subsets.

Inspecting the layout of the stations data with `View(ny_stations)` you will notice that the top line is the header and contains column (variable) names. The `glimpse()` function can be used to quickly describe a data frame's dimensions. We have 500,000 trip observations in `ny_trips` and 11 variables; the 500,000 represents a random sample of c.1.9 million trips recorded in June 2020. The function also prints out the object type for each of these variables, with the variables either of type `int`, `chr` or `dbl`.

In this case the assignment needs correcting. `start_time` and `stop_time` may be better represented in `date-time` format; the station identifier variables (e.g. `start_station_id`) are more efficient when converted to `int` types; and the geographic coordinates, currently stored as text strings (`chr`), are better converted as floating points (`dbl`) or `POINT` geometry types (Pebesma 2018). In the `02-template.qmd` file are code chunks for doing these conversions. There are some slightly more involved data transform procedures in this code, which you may wish to ignore at this stage.

```
glimpse(ny_trips)
## Rows: 500,000
```

---

[2]https://github.com/vis4sds/data

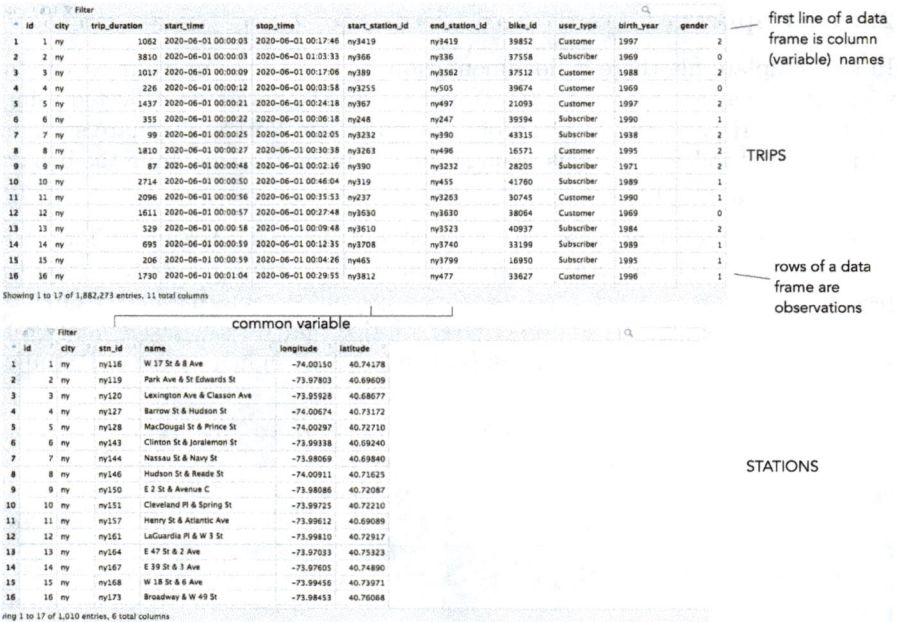

**Figure 2.1:** `ny_trips` and `ny_stations` as they appear when calling `View()`.

```
## Columns: 11
## $ id              <int> 1, 2, 3, 4, 5, 6, 7...
## $ city            <chr> "ny", "ny", "ny", "n.."
## $ trip_duration   <dbl> 1062, 3810, 1017, 226..
## $ start_time      <chr> "2020-06-01 00:00:03", ...
## $ stop_time       <chr> "2020-06-01 00:17:46", ...
## $ start_station_id <chr> "ny3419", "ny366", "ny389", ...
## $ end_station_id  <chr> "ny3419", "ny336", "ny3562", ...
## $ bike_id         <chr> "39852", "37558", "37512", ...
## $ user_type       <chr> "Customer", "Subscriber", ...
## $ birth_year      <chr> "1997", "1969", "1988", ...
## $ gender          <dbl> 2, 0, 2, 0, 2, 1, 2, 2, ...


glimpse(ny_stations)
## Rows: 1,010
## Columns: 6
## $ id       <int> 1, 2, 3, 4, 5, 6, 7, ...
## $ city     <chr> "ny", "ny", "ny", "ny", "ny", "n..."
## $ stn_id   <chr> "ny116", "ny119", "ny120", "ny127", "n..."
```

**Table 2.2:** dplyr functions (verbs) for manipulating data frames.

| function | description |
|---|---|
| `filter()` | Picks rows (observations) if their values match a specified criteria |
| `arrange()` | Reorders rows (observations) based on their values |
| `select()` | Picks a subset of columns (variables) by name (or name characteristics) |
| `rename()` | Changes the name of columns in the data frame |
| `mutate()` | Adds new columns |
| `group_by()` | Chunks the dataset into groups for grouped operations |
| `summarise()` | Calculates single-row (non-grouped) or multiple-row (if grouped) summary values |
| ... | |

```
## $ name      <chr> "W 17 St & 8 Ave", "Park Ave", "B...."
## $ longitude <chr> "-74.00149746", "-73.97803415", "..."
## $ latitude  <chr> "40.74177603", "40.69608941", "..."
```

## 2.3.2 Manipulate

### Manipulate with dplyr and pipes (|>)

dplyr is the foundational package of the tidyverse. It provides a *grammar of data manipulation*, with access to functions that can be variously combined to support most data processing and manipulation tasks. Once familiar with dplyr functions, you will find yourself generating analysis templates to re-use whenever working on a dataset.

All dplyr functions operate in a consistent way:

1. Start with a data frame.
2. Pass arguments to a function performing some updates to the data frame.
3. Return the updated data frame.

So every dplyr function expects a data frame and will always return a data frame.

dplyr functions are designed to be chained together, and this chaining of functions can be achieved using the *pipe* operator (|>). Pipes are mechanisms for passing information in a program. They take the output of a set of code (a dplyr specification) and make it the input of the next set (another dplyr specification). Pipes can be easily applied to dplyr functions and the functions of all packages that form the tidyverse. We mentioned in Chapter 1 that ggplot2 provides a framework for specifying a *layered grammar of graphics* (more on this in Chapter 3). Together with the pipe operator, `dplyr` supports a *layered grammar of data manipulation*. You will see this throughout the book as we develop and re-use code templates for performing some data manipulation that is then piped to a ggplot2 specification for visual analysis.

### `count()` and `summarise()` over rows

Let's combine some `dplyr` functions to generate statistical summaries of the New York bikeshare data. First we'll count the number of trips made in June 2020 by `user_type`, a variable distinguishing casual users from those formally registered to use the system (`Customer` vs. `Subscriber` cyclists). `dplyr` has a convenience function for counting, so we could run the code below, also in the `02-template.qmd` for this chapter.

```
# Take the ny_trips data frame.
ny_trips |>
  # Run the count function and sort the result.
  count(user_type, sort=TRUE)
##     user_type      n
## 1 Subscriber 347204
## 2   Customer 152796
```

There are a few things happening in the `count()` function. It takes the `usr_type` variable from `ny_trips`, organises or *groups* the rows in the data frame according to its values (`Customer` | `Subscriber`), counts the rows and then orders the summarised output descending on the counts.

Often you will want to do more than simply counting, and you may also want to be more explicit in the way the data frame is grouped for computation. A common workflow is to combine `group_by()` and `summarise()`, and in this case `arrange()` to replicate the `count()` example.

```
# Take the ny_trips data frame.
ny_trips |>
  # Group by user_type.
  group_by(user_type) |>
    # Count the number of observations per group.
    summarise(count=n()) |>
    # Arrange the grouped and summarised (collapsed) rows in count.
    arrange(desc(count))
## # A tibble: 2 × 2
##   user_type    count
##   <chr>        <int>
## 1 Subscriber 347204
## 2 Customer   152796
```

In `ny_trips` there is a variable measuring trip duration in seconds (`trip_duration`). It may be instructive to calculate some summary statistics to see how trip duration varies between these groups. The code below uses `group_by()`, `summarise()` and `arrange()` in exactly the same way, but with the addition of other aggregate functions summarises the `trip_duration` variable according to central tendency (mean and standard deviation) and by `user_type`.

```r
# Take the ny_trips data frame.
ny_trips |>
  mutate(trip_duration=trip_duration/60) |>
  # Group by user type.
  group_by(user_type) |>
  # Summarise over the grouped rows,
  # generate a new variable for each type of summary.
  summarise(
    count=n(),
    avg_duration=mean(trip_duration),
    median_duration=median(trip_duration),
    sd_duration=sd(trip_duration),
    min_duration=min(trip_duration),
    max_duration=max(trip_duration)
  ) |>
  # Arrange on the count variable.
  arrange(desc(count))

## # A tibble: 2 × 7
##   user_type    count avg_dur med_dur sd_dur min_dur max_dur
##   <chr>        <int>   <dbl>   <dbl>  <dbl>   <dbl>   <dbl>
## 1 Subscriber  347204    20.3    14.4   116.    1.02  33090.
## 2 Customer    152796    43.3    23.1   383.    1.02  46982.
## # ... with abbreviated variable names
```

As each line is commented you hopefully get a sense of what is happening in the code above. Since `dplyr` functions read like verbs, and code is executed sequentially, it greatly helps to organise `dplyr` code such that each new verb (function call) occupies a single line, separated with a pipe (`|>`). Once you are familiar with `dplyr`, it becomes very easy to read, write, re-use and share code in this way.

> **i** On pipes `|>`
>
> Remembering that pipes take the output of a set of code and make it the input of the next set, separate lines are used for each call to the pipe operator. This is good practice for supporting readability of your code,

**Table 2.3:** A breakdown of aggregate functions commonly used with `summarise()`.

| Function | Description |
|---|---|
| `n()` | Counts the number of observations |
| `n_distinct(var)` | Counts the number of unique observations |
| `sum(var)` | Sums the values of observations |
| `max(var)\|min(var)` | Finds the min\|max values of observations |
| `mean(var)\|median(var)\| ...` | Calculates central tendency of observations |
| `...` | Many more |

and for debugging and learning how your data is affected by each line. Especially if `dplyr` is new to you, we recommend you run each code line separated by a pipe (`|>`) in the Console and observe how the dataset is changed.

### Manipulate dates with `lubridate`

Let's continue this investigation of trips by user type by profiling how usage varies over time. To do this we will need to work with `date-time` variables. The `lubridate` package provides various convenience functions for this.

In the code block below we extract the *day of week* and *hour of day* from the `start_time` variable using `lubridate`'s day accessor functions. Documentation on these can be accessed in the usual way (`?<function-name>`). Next we count the number of trips made by hour of day, day of week and user type. The summarised data frame will be re-used several times in our analysis, so we store it as an object with a suitable name (`ny_temporal`) using the assignment operator.

```r
# Create an hour of day and day of week summary by user type
# and assign it the name "ny_temporal".
ny_temporal <- ny_trips |>
  mutate(
    # Create a new column to identify dow.
    day=wday(start_time, label=TRUE),
    # Create a new column to identify hod.
    hour=hour(start_time)) |>
  # Group by day, hour, user_type.
  group_by(user_type, day, hour) |>
  # Count the grouped rows.
  summarise(count=n()) |>
  ungroup()
```

> **i** Keeping track of derived data
>
> Whether or not to store derived data frames, like the newly assigned `ny_temporal`, in a session is not an easy decision. You should avoid cluttering the Environment pane with many data objects. Often when generating charts it is necessary to create these sorts of derived tables as input data to ggplot2. Adding derived data frames to the RStudio Environment pane each time an exploratory plot is created risks an unhelpfully large number of such tables. A general rule: if the derived data frame is to be used more than three times in a data analysis or is computationally intensive, create and assign it (`<-`) as a named object.

In Figure 2.2 the newly derived data are plotted. Code for creating this data graphic is below. The plot demonstrates a familiar weekday-weekend pattern of usage. Trip frequencies peak in the morning and evening rush hours during weekdays and mid/late-morning and afternoon during weekends, with the weekday afternoon peak much larger than the morning peak. There are obvious differences in the type of trips made by subscribers versus customers – the temporal signature for subscribers appears to match more closely what one would expect of commuting behaviour. That this pattern exists is notable when remembering the data are from June 2020, a time when New York residents were emerging from lockdown. It would be instructive to compare with data from a non-Covid year. The fact that bikeshare is recorded continuously, in contrast to actively-collected survey data, makes this sort of behavioural change analysis possible.

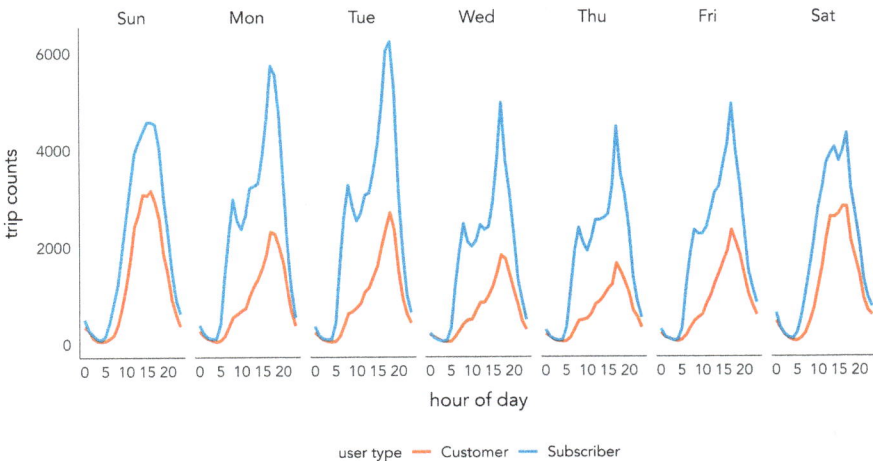

**Figure 2.2:** Citibike trips made by hour of day and day and week, differentiated by customer type.

```
# Calculate trips by hour of day, day and customer type.
ny_trips |>
  mutate(
    day=wday(start_time, label=TRUE),
    hour=hour(start_time)) |>
  group_by(user_type, day, hour) |>
  summarise(count=n()) |>
  ungroup() |>
  # Pipe to ggplot2 for plotting.
  ggplot(aes(x=hour, y=count)) +
  geom_line(aes(colour=user_type), size=1) +
  scale_colour_manual(values=c("#e31a1c", "#1f78b4")) +
  facet_wrap(~day, nrow=1)+
  labs(x="hour of day", y="trip counts", colour="user type")
```

## Relate tables with `join()`

Trip distance is not recorded directly in the `ny_trips` table, but may
be important for profiling usage behaviour. Since `ny_stations` contains
coordinates corresponding to station locations, distances can be calculated by linking these station coordinates to the origin and destination
stations recorded in `ny_trips`. To relate the two tables we need to specify
a join between them.

A sensible approach is to:

1. Select all uniquely cycled trip pairs (origin-destination pairs) that
   appear in the `ny_trips` table.
2. Bring in the corresponding coordinate pairs representing the origin
   and destination stations by joining on the `ny_stations` table.
3. Calculate the distance between the coordinate pairs representing
   the origin and destination.

The code below is one way of achieving this.

```
# Take the ny_trips data frame.
od_pairs <- ny_trips |>
  # Select trip origin and destination (OD) station columns
  # and extract unique OD pairs.
  select(start_station_id, end_station_id) |> unique() |>
  # Select lon, lat columns from ny_stations and join on origin column.
  left_join(
    ny_stations |> select(stn_id, longitude, latitude),
    by=c("start_station_id"="stn_id")
```

**Table 2.4:** A breakdown of `dplyr` join functions.

| | |
|---|---|
| `left_join()` | all rows from table x |
| `right_join()` | all rows from table y |
| `full_join()` | all rows from both table x and y |
| `semi_join()` | all rows from table x where there are matching values in table y, keeping just columns from table x |
| `inner_join()` | all rows from table x where there are matching values in table y, returning all combinations where there are multiple matches |
| `anti_join` | all rows from table x where there are not matching values in table y, never duplicating rows of table x |

```
  ) |>
# Rename new lon, lat columns and associate with origin station.
rename(o_lon=longitude, o_lat=latitude) |>
# Select lon, lat columns from ny_stations and join
# on destination column.
left_join(
   ny_stations |> select(stn_id, longitude, latitude),
   by=c("end_station_id"="stn_id")
   ) |>
# Rename new lon, lat columns and associate with destination station.
rename(d_lon=longitude, d_lat=latitude) |>
# Compute distance calculation on each row (od_pair).
rowwise() |>
# Calculate distance and express in kms.
mutate(
   dist=
   geosphere::distHaversine(c(o_lat, o_lon), c(d_lat, d_lon))/1000
   ) |>
ungroup()
```

Some new functions are introduced: `select()` to pick or drop variables, `rename()` to rename variables and a convenience function for calculating straight line distances from polar coordinates, `distHaversine()`. The key function to emphasise is the `left_join()`. If you've worked with relational databases, `dplyr`'s join functions will be familiar to you. In a `left_join` all the values from the first (left-most) table are retained, `ny_trips` in this case, and variables from the table on the right , `ny_stations`, are added. We specify the variable on which tables should be joined with the `by=` parameter, `station_id` in this case. If there is a `station_id` in `ny_trips` that doesn't exist in `ny_stations` then corresponding cells are filled out with NA.

From the newly created distance variable we can calculate the average (mean) trip distance for the 500,000 sampled trips – 1.6km. This might seem very short, but remember that these are straight-line distances between pairs of docking stations. Ideally we would calculate distances derived from cycle trajectories. A separate reason, discovered when generating a histogram on the `dist` variable, is that there are a large number of trips that start and end at the same docking station. These might be unsuccessful hires – people failing to undock a bike for example. We could investigate this further by paying attention to the docking stations at which same origin-destination trips occur, as in the code block below.

```
ny_trips |>
  filter(start_station_id==end_station_id) |>
  group_by(start_station_id) |> summarise(count=n()) |>
  left_join(
    ny_stations |> select(stn_id, name),
    by=c("start_station_id"="stn_id")
    ) |>
  arrange(desc(count))
## # A tibble: 958 x 3
##    start_station_id count name
##    <chr>            <int> <chr>
##  1 ny3423            2017 West Drive & Prospect Park West
##  2 ny3881            1263 12 Ave & W 125 St
##  3 ny514             1024 12 Ave & W 40 St
##  4 ny3349             978 Grand Army Plaza & Plaza St West
##  5 ny3992             964 W 169 St & Fort Washington Ave
##  6 ny3374             860 Central Park North & Adam Clayton Powell
##  7 ny3782             837 Brooklyn Bridge Park - Pier 2
##  8 ny3599             829 Franklin Ave & Empire Blvd
##  9 ny3521             793 Lenox Ave & W 111 St
## 10 ny2006             782 Central Park S & 6 Ave
## # ... with 948 more rows
```

The top 10 docking stations are either in parks, near parks or located along the river. This, coupled with the fact that same origin-destination trips occur in much greater relative number for casual users (`Customer`), associated with discretionary leisure-oriented cycling, than regular users (`Subscriber`) is further evidence that these are valid trips. Note also the different shapes in the distribution of distances for trips cycled by subscribers and customers (Figure 2.3), again suggesting these groups may use Citibike in different ways. Code for creating Figure 2.3 appears below the graphic.

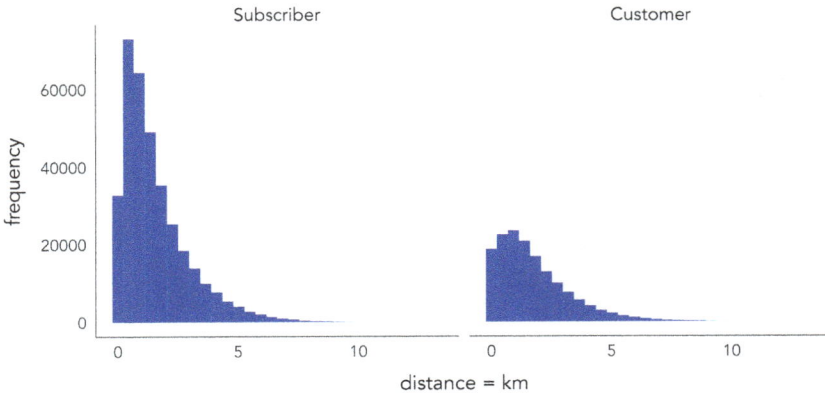

**Figure 2.3:** Citibike trip distances (straight-line km) for Subscriber and Customer cyclists.

```
# Plot faceted histograms.
ny_trips |>
  # Cast as factor variable to draw Subscriber plot facet first.
  mutate(
    user_type=factor(user_type, levels=c("Subscriber", "Customer"))
  ) |>
  ggplot(aes(dist)) +
  geom_histogram() +
  facet_wrap(~user_type)+
  labs(x="distance = km", y="frequency")
```

## Write functions of your own

There may be times when you need to create functions of your own. Most often this is when you find yourself copy-pasting the same chunks of code with minimal adaptation.

Functions have three key characteristics:

1.  They are (usually) named – the name should be expressive and communicate what the function does.
2.  They have brackets `<function-name()>` usually containing arguments – inputs, which determine what the function does and returns.
3.  Immediately followed by `<function-name()>` are curly brackets (`{}`) used to contain the body – code that performs a distinct task, described by the function's name.

Effective functions are short and perform single, discrete operations.

You will recall that in the `ny_trips` table there is a variable called `birth_year`. From this we can derive cyclists' approximate age in years. Below is a function called `get_age()`. The function expects two arguments: `yob` – a year of birth as type `chr`; and `yref` – a reference year. In the body, `lubridate`'s `as.period()` function is used to calculate the time in years that elapsed between these dates.

```
# get_age() depends on lubridate.
library(lubridate)

# Calculate time elapsed between two dates in years (age).
# yob : datetime object recording birth year.
# yref : datetime object recording reference year.
get_age <- function(yob, yref) {
    period <- as.period(interval(yob, yref),unit = "year")
    return(period$year)
}

ny_trips <- ny_trips |>
  # Calculate age from birth_date.
  mutate(
    age=get_age(
      as.POSIXct(birth_year, format="%Y"),
      as.POSIXct("2020", format="%Y")
    )
  )
```

We can use the two new derived variables – distance travelled and age – in our analysis. In Figure 2.4, we explore how approximate travel speeds vary by age, trip distance and customer type. Again the average speed calculation should be treated cautiously as it is based on straight line distances, and it is likely that this will vary depending on whether the trip is made for 'utilitarian' or 'leisure' purposes. Additionally, due to the heavy subsetting data become a little volatile for certain age groups, and so the age variable is aggregated into 5-year bands.

There are some notable patterns in Figure 2.4. Subscribers make faster trips than do customers, although this gap narrows as trip distance increases. Trips with a straight-line distance of 4.5km are non-trivial and so may be classed as utilitarian even for non-regular customers. There is a very slight effect of decreasing trip speed by age cycled for the longer trips. The volatility in the older age groups for trips >4.5km suggests more data, and a more involved analysis, is required to confidently establish this. For example, it may be that the comparatively rare occurrence of trips in the 65-70 age group is made by only a small subset of cyclists; with a larger dataset we may expect a regression to the mean effect that negates noise caused by outlier individuals.

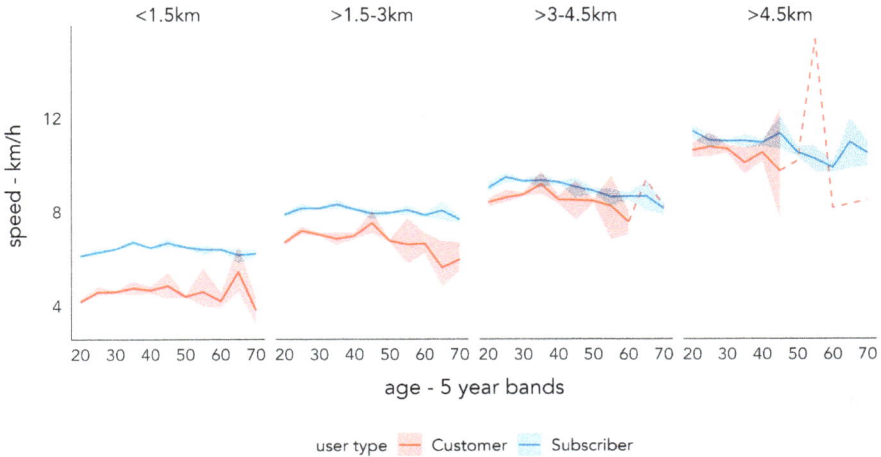

**Figure 2.4:** Citibike average trip speeds (approximate) by age, customer type and straight-line trip distance.

```
# Generate staged dataset for plotting.
# Filter weekday-only trips, less than 1hr in duration,
# exclude trips that start and end at same docking station,
# and generate binned age and distance variables.
temp_data <- ny_trips |>
  mutate(
    day=wday(start_time, label=TRUE),
    is_weekday=as.numeric(!day %in% c("Sat", "Sun"))
    ) |>
  filter(
    is_weekday==1,
    start_station_id!=end_station_id,
    duration_minutes<=60,
    between(age, 16, 74),
    dist>.5
    ) |>
  mutate(
    dist_bands=case_when(
      dist < 1.5 ~ "<1.5km",
      dist < 3 ~ ">1.5-3km",
      dist < 4.5 ~ ">3-4.5km",
```

```
    TRUE ~ ">4.5km"),
  age_band=if_else(age %% 10 > 4, ceiling(age/5)*5,
    floor(age/5)*5),
  speed=dist/(duration_minutes/60)
) |>
group_by(user_type, age_band, dist_bands) |>
  summarise(avg_speed=mean(speed), sample_size=n(),
    std=sd(speed)) |>
ungroup()


# Plot.
temp_data |>
  ggplot(aes(x=age_band, y=avg_speed))+
  geom_line(aes(colour=user_type)) +
  scale_colour_manual(values=c("#e31a1c", "#1f78b4")) +
  scale_fill_manual(values=c("#e31a1c", "#1f78b4")) +
  facet_wrap(~dist_bands, nrow=1) +
  labs(
    x="age", y="speed - km/h ", fill="user type",
    colour="user type"
  )


# Remove staging dataset.
rm(temp_data)
```

### 2.3.3   Tidy

The `ny_trips` and `ny_stations` data already comply with the rules for tidy data
(Wickham 2014). Each row in `ny_trips` is a distinct trip and each row in
`ny_stations` a distinct station. However, it is common to encounter datasets
that are untidy and must be reshaped. In the book's data repository are
two examples: `ny_spread_rows` and `ny_spread_columns`. `ny_spread_rows` is so-called
because the variable `summary_type` is spread across the rows (observations);
`ny_spread_columns` because multiple variables are stored in single columns – the
`dist_weekday, duration_weekday` columns.

```
ny_spread_rows
## # A tibble: 411,032 × 6
##    o_station d_station wkday   count summary_type value
##        <dbl>     <dbl> <chr>   <dbl> <chr>        <dbl>
## 1         72       116 weekend     1 dist          1.15
## 2         72       116 weekend     1 duration     18.2
## 3         72       127 weekend     4 dist          7.18
```

```
## 4          72        127 weekend    4 duration   122.
## 5          72        146 weekend    4 dist         9.21
## 6          72        146 weekend    4 duration   122.
## 7          72        164 weekend    1 dist         2.66
## 8          72        164 weekend    1 duration    12.5
## 9          72        173 weekend    2 dist         2.13
## 10         72        173 weekend    2 duration    43.6
## # ... with 411,022 more rows
```

```
ny_spread_columns
## # A tibble: 156,449 × 8
##    o_stat d_stat ct_wend ct_wdy dst_wnd dst_wdy dur_wnd dur_wdy
##     <dbl>  <dbl>   <dbl>  <dbl>   <dbl>   <dbl>   <dbl>   <dbl>
## 1     72    116       1      3    1.15    3.45    18.2    49.9
## 2     72    127       4      4    7.18    7.18   122.    101.
## 3     72    146       4      2    9.21    4.61   122.     64.1
## 4     72    164       1      1    2.66    2.66    12.5    43.2
## 5     72    173       2     13    2.13   13.9     43.6   189.
## 6     72    195       1      4    2.56   10.2     24.7    98.3
## 7     72    212       3      3    4.83    4.83    40.3    54.0
## 8     72    223       1     NA    1.13      NA    21.1      NA
## 9     72    228       2      1    4.97    2.49    30.2    13.6
## 10    72    229       1     NA    1.22      NA    39.2      NA
## # ... with 156,439 more rows, and abbreviated variable names
```

To re-organise the table in tidy form, we should identify what constitutes a distinct observation – an origin-destination pair summarising counts, distances and durations of trips that occur during the weekday or weekend. From here, the table's variables are:

- o_station: station id of the origin station
- d_station: station id of the destination station
- wkday: trip occurs on weekday or weekend
- count: count of trips for observation type
- dist: total straight-line distance in km (cumulative) of trips for observation type
- duration: total duration in minutes (cumulative) of trips for observation type

There are two functions for reshaping untidy data, from the tidyr package: pivot_longer() and pivot_wider(). pivot_longer() is used to tidy data in which observations are spread across columns; pivot_wider() to tidy data in which variables are spread across rows. The functions are especially useful in visual data analysis to fix messy data, but also to flexibly reshape data supplied to ggplot2 specifications (more on this in Chapters 3 and 4).

To fix `ny_spread_rows`, we use `pivot_wider()` and pass to the function's arguments the name of the problematic column and the column containing values used to populate the newly created columns.

```
ny_spread_rows |>
   pivot_wider(names_from=summary_type, values_from=value)
```

```
## # A tibble: 205,516 × 6
##     o_station d_station wkday    count  dist duration
##         <dbl>     <dbl> <chr>    <dbl> <dbl>    <dbl>
## 1          72       116 weekend      1  1.15     18.2
## 2          72       127 weekend      4  7.18    122.
## 3          72       146 weekend      4  9.21    122.
## 4          72       164 weekend      1  2.66     12.5
## 5          72       173 weekend      2  2.13     43.6
## 6          72       195 weekend      1  2.56     24.7
## 7          72       212 weekend      3  4.83     40.3
## 8          72       223 weekend      1  1.13     21.1
## 9          72       228 weekend      2  4.97     30.2
## 10         72       229 weekend      1  1.22     39.2
## # ... with 205,506 more rows
```

To fix `ny_spread_columns` requires a little more thought. First we `pivot_longer()` on columns that are muddled with multiple variables. This results in a long and thin dataset similar to `ny_spread_rows` – each row is the origin-destination pair with either a count, distance or duration recorded for trips occurring on weekends or weekdays. The muddled variables, for example `dist_weekend` `duration_weekday`, now appear in the rows of a new column with the default title `name`. This column is separated on the _ mark to create two new columns, `summary_type` and `wkday`, used in `pivot_wider()`.

```
ny_spread_columns |>
  pivot_longer(cols = count_weekend:duration_weekday) |>
  separate(col = name, into = c("summary_type", "wkday"), sep = "_") |>
  pivot_wider(names_from = summary_type, values_from = value)
```

```
## # A tibble: 312,898 × 6
##     o_station d_station wkday    count  dist duration
##         <dbl>     <dbl> <chr>    <dbl> <dbl>    <dbl>
## 1          72       116 weekend      1  1.15     18.2
## 2          72       116 weekday      3  3.45     49.9
## 3          72       127 weekend      4  7.18    122.
## 4          72       127 weekday      4  7.18    101.
## 5          72       146 weekend      4  9.21    122.
```

```
## 6            72        146 weekday      2  4.61      64.1
## 7            72        164 weekend      1  2.66      12.5
## 8            72        164 weekday      1  2.66      43.2
## 9            72        173 weekend      2  2.13      43.6
## 10           72        173 weekday     13 13.9       189.
## # ... with 312,888 more rows
```

---

**Task 2**

Figure 2.2 uses a derived dataset that summarises trip counts by `user_type`
and `day_of_week`. This dataset is created in the template file for the chapter
and is named `ny_temporal`. Each observation is a trip count on a day of
week, hour of day and for a given user type (`Customer` or `Subscriber`).

```
ny_temporal <- ny_trips |>
  mutate(
    day=wday(start_time, label=TRUE),
    hour=hour(start_time)) |>
  group_by(user_type, day, hour) |>
  summarise(count=n()) |>
  ungroup()
```

To explore whether customers and subscribers have different usage behaviours, we calculate the proportion of trips made by day of week for
these two user groups. Customers, as expected, contribute a greater
relative number of trips on weekends than do subscribers.

```
# # A tibble: 7 × 3
#     day    Customer Subscriber
#     <ord>     <dbl>      <dbl>
# 1 Sun       0.198      0.144
# 2 Mon       0.137      0.163
# 3 Tue       0.144      0.172
# 4 Wed       0.104      0.125
# 5 Thu       0.0973     0.122
# 6 Fri       0.135      0.138
# 7 Sat       0.185      0.136
```

Can you write some `dplyr` code to generate such a summary? There are
several possible approaches, but you will need to think about which variables to `group_by()` and `summarise()` over, and you may need to `pivot_wider()`
your dataset in order to compare the user types side-by-side.

---

## 2.4   Conclusions

Developing the vocabulary and technical skills to systematically describe and organise data is crucial to modern data analysis. This chapter has covered the fundamentals: that data consist of *observations* and *variables* of different types (Stevens 1946) and that in order to work effectively with datasets, these data should be organised such that they are *tidy* (Wickham 2014). Most of the chapter content was dedicated to the techniques that enable these concepts to be operationalised. We covered how to download, transform and reshape a reasonably large dataset from New York's Citibike system. In doing so, we generated insights that might inform further data collection and analysis activity. In the next chapter we will extend this conceptual and technical knowledge as we introduce the fundamentals of visual data analysis and ggplot2's *grammar of graphics*.

## 2.5   Further Reading

There are many accessible introductions to the tidyverse for modern data analysis. Two excellent resources:

- Wickham, H., Çetinkaya-Rundel, M., Grolemund, G. 2023, "R for Data Science, 2nd Edition", Sebastopol, CA: *O'Reilly*.
  - Chapter 3.
- Ismay, C. and Kim, A. 2020. "Statistical Inference via Data Science: A ModernDive into R and the Tidyverse", New York, NY: *CRC Press*. doi: 10.1201/9780367409913.
  - Chapters 3, 4.

Hadley Wickham's original paper on Tidy Data:

- Wickham, H. 2010. "Tidy Data" *Journal of Statistical Software*, 59(10): 1–23. doi: 10.18637/jss.v059.i10.

# 3

## *Visualization Fundamentals*

By the end of this chapter you should gain the following knowledge and practical skills.

**Knowledge outcomes**

☐ Recognise the characteristics of effective data graphics.
☐ Understand that there is a *grammar* of graphics, and that this grammar underpins modern visualization toolkits (ggplot2, vega-lite and Tableau).
☐ Appreciate how visual channels and knowledge of their encoding effectiveness can be used to design and evaluate data graphics.

**Skills outcomes**

☐ Write ggplot2 code to generate statistical graphics (histograms, bar charts, scatterplots, choropleth maps).
☐ Update that code to *layer* graphics with multiple variables.
☐ Write code to manipulate the order and colour of data items in statistical graphics.
☐ Advanced: create glyphmaps in ggplot2 by writing code that works on shape primitives.
☐ (Very) Advanced: create dot-density maps in ggplot2 using re-sampling and functional programming.

## 3.1 Introduction

This chapter outlines the fundamentals of visualization design. It offers a position on what effective data graphics should do, before discussing the processes that take place when creating data graphics. A framework – a vocabulary and grammar – for supporting this process is presented which, combined with established knowledge on visual perception, helps describe, evaluate and create effective data graphics. Talking about a vocabulary and grammar of data and graphics may sound somewhat abstract. However, through an analysis of 2019 General Election results data, the chapter will demonstrate how these concepts are fundamental to visual data analysis.

## 3.2  Concepts

### 3.2.1  Effective data graphics

Data graphics take numerous forms and are used in many different ways by scientists, journalists, designers and many more. While the intentions of those producing them may vary, data graphics that are effective generally have the following characteristics:

- Expose complex structure, connections and comparisons that could not be achieved easily via other means;
- Are data rich, presenting many numbers in a small space;
- Reveal patterns at several levels of detail, from broad overview to fine structure;
- Are concise, emphasising dimensions of a dataset without extraneous details;
- Generate an aesthetic response, encouraging people to engage with the data or question.

Consider the data graphic in Figure 3.1, which presents an analysis of the 2016 US Presidential Election, or the *Peaks and Valleys of Trump and Clinton's Support*. The map is reproduced from an article in The Washington Post (Gamio and Keating 2016). Included in the bottom margin is a choropleth map coloured according to party majority, more standard practice for reporting county-level voting. Gamio and Keating's (2016) graphic is clearly *data rich*, encoding many more data items than does the standard choropleth. It is not simply the data density that makes the graphic successful, however. There are careful design choices that help *support comparison* and emphasise *complex structure*. By varying the height of triangles according to the number of votes cast, the thickness according to whether or not the result for Trump/Clinton was a landslide and rotating the map 90 degrees, the very obvious differences between metropolitan, densely populated coastal counties that voted emphatically for Clinton and the vast number of suburban, provincial town and rural counties (everywhere else) that voted for Trump, are exposed.

## 3.2.2   Grammar of Graphics

---

"Data graphics visually display measured quantities by means of
the combined use of points, lines, a coordinate system, numbers,
symbols, words, shading, and color."

Tufte (1983)

---

So the Washington Post graphic demonstrates a judicious *mapping* of data to
visuals, underpinned by a close appreciation of the analysis context. The act
of carefully considering how best to leverage visual systems given the available
data and analysis priorities is key to designing effective data graphics. Leland
Wilkinson's *Grammar of Graphics* (1999) captures this process of turning
data into visuals. Wilkinson's (1999) thesis is that graphics can be described
in a consistent way according to their structure and composition. This has
obvious benefits for building visualization toolkits. If different chart types and
combinations can be reduced to a common vocabulary and grammar, then
the process of designing and *generating* graphics of different types can be
systematised.

Wilkinson's (1999) grammar separates the construction of data graphics into a
series of components. Below are the components of the *Layered Grammar of
Graphics* on which ggplot2 is based (Wickham 2010), adapted from Wilkinson's
(1999) original work. The components in Figure 3.2 are used to assemble
ggplot2 specifications. Those to highlight at this stage are in **emphasis**: the
data containing the variables of interest, the marks used to represent data
and the visual channels through which variables are encoded.

To demonstrate this, let's generate some scatterplots based on the 2019 General
Election data. Two variables worth exploring for association here are: `con_1719`,
the change in Conservative vote share by constituency between 2017-2019, and
`leave_hanretty`, the size of the Leave vote in the 2016 EU referendum, estimated
at Parliamentary Constituency level (via Hanretty 2017).

In Figure 3.3 are three plots, accompanied by ggplot2 specifications used to
generate them. Reading the graphics and the associated code, you should get
a feel for how ggplot2 specifications are constructed:

1.   Start with a data frame, in this case 2019 General Election results for
     UK Parliamentary Constituencies. The data are passed to ggplot2
     (`ggplot()`) using the pipe operator (`|>`). Also at this stage, we consider
     the variables to encode and their measurement type – both `con_1719`
     and `leave_hanretty` are `ratio` scale variables.

## THE PEAKS & VALLEYS OF TRUMP-CLINTON SUPPORT
### CLINTON WON IN URBAN COUNTIES, WHILE TRUMP WON EVERYWHERE ELSE

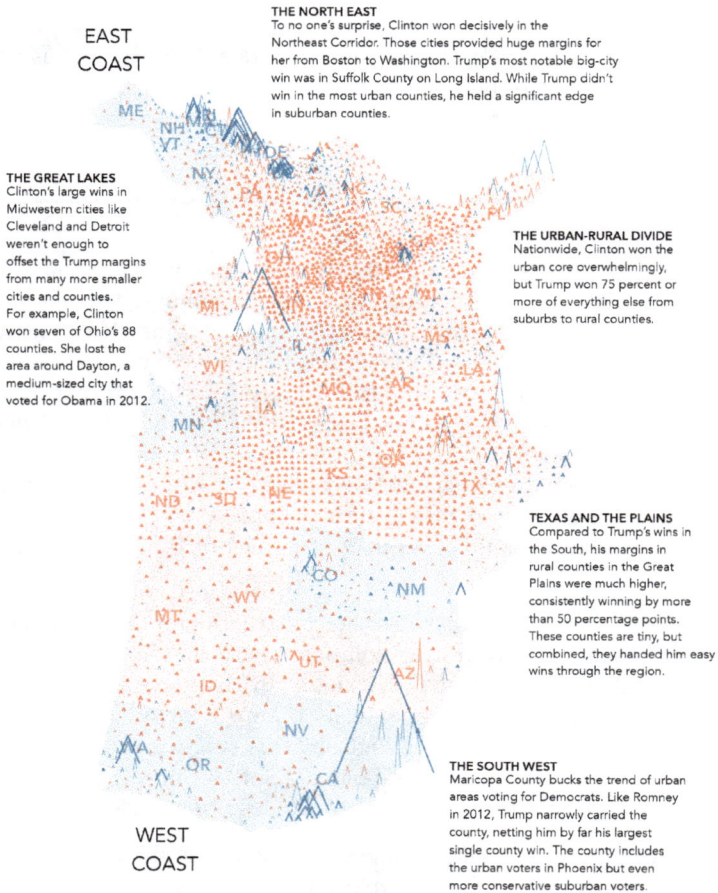

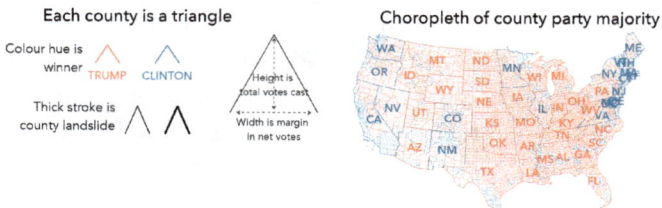

EAST
COAST

**THE NORTH EAST**
To no one's surprise, Clinton won decisively in the Northeast Corridor. Those cities provided huge margins for her from Boston to Washington. Trump's most notable big-city win was in Suffolk County on Long Island. While Trump didn't win in the most urban counties, he held a significant edge in suburban counties.

**THE GREAT LAKES**
Clinton's large wins in Midwestern cities like Cleveland and Detroit weren't enough to offset the Trump margins from many more smaller cities and counties. For example, Clinton won seven of Ohio's 88 counties. She lost the area around Dayton, a medium-sized city that voted for Obama in 2012.

**THE URBAN-RURAL DIVIDE**
Nationwide, Clinton won the urban core overwhelmingly, but Trump won 75 percent or more of everything else from suburbs to rural counties.

**TEXAS AND THE PLAINS**
Compared to Trump's wins in the South, his margins in rural counties in the Great Plains were much higher, consistently winning by more than 50 percentage points. These counties are tiny, but combined, they handed him easy wins through the region.

**THE SOUTH WEST**
Maricopa County bucks the trend of urban areas voting for Democrats. Like Romney in 2012, Trump narrowly carried the county, netting him by far his largest single county win. The county includes the urban voters in Phoenix but even more conservative suburban voters.

WEST
COAST

Each county is a triangle
Colour hue is winner — TRUMP / CLINTON
Thick stroke is county landslide
Height is total votes cast
Width is margin in net votes

Choropleth of county party majority

**Figure 3.1:** Map of 2016 US presidential election results. Note that for copyright reasons this is a re-implementation in ggplot2 of Gamio and Keating's (2016) original, which appeared in The Washington Post.

| Required | | Not required (sensible defaults) | |
|---|---|---|---|
| Data | Variables and data types to be represented<br>data ▷ <some dplyr> ▷ ... | Statistics | stat_count(), stat_density(), stat_quantile(), ... |
| | | Scale | scale_*_continuous(), scale_*_datetime(), ... |
| Aesthetics | Visual channel that mapped to variable in data<br>aes(x,y, colour, shape, size, alpha, angle, ...) | Coordinates | coord_cartesian(), coord_polar(), coord_map(), ... |
| | | Facets | facet_wrap(), facet_grid(), ... |
| Geom | Shapes, or marks, used to present data<br>geom_point(), geom_line(), geom_bar(), ... | Themes | theme_bw(), theme_classic(), theme_minimal(), ... |

**Figure 3.2:** Components of Wickham's (2010) Layered Grammar of Graphics.

2. Next is the encoding (`mapping=aes()`), which determines how the data are to be mapped to visual **channels**. In a scatterplot, horizontal and vertical position varies in a meaningful way, in response to the values of a dataset. Here the values of `leave_hanretty` are mapped along the x-axis, and the values of `con_1719` are mapped along the y-axis.

3. Finally, we represent individual data items with **marks** using the `geom_point()` geometry.

In the middle plot, the grammar is updated such that the points are coloured according to `winning_party`, a variable of type categorical `nominal`. In the bottom plot constituencies that flipped from Labour-to-Conservative between 2017-19 are emphasised by varying the `shape` (filled and not filled) and transparency (`alpha`) of points.

### 3.2.3 Marks and visual channels

In our descriptions **marks** was used as an alternative term for *geometry* and visual encoding **channels** as an alternative for *aesthetics*. We also paid special attention to the **data types** that were encoded. *Marks* are graphical elements such as *bars*, *lines*, *points* and *ellipses* that can be used to represent data items. In ggplot2 marks are accessed through the function layers prefaced with `geom_*()`. Visual *channels* are attributes such as *colour*, *size* and *position* that, when mapped to data, affect the appearance of marks in response to the values of a dataset. These attributes are controlled via the `aes()` (aesthetics) function in ggplot2.

*Marks* and *channels* are terms used routinely in Information Visualization, an academic discipline devoted to the study of data graphics, and most notably by Tamara Munzner (2014) in her textbook *Visualization Analysis and Design*. Munzner's (2014) work synthesises over foundational research in Information Visualization and Cognitive Science testing how effective different visual channels are at supporting specific tasks. Figure 3.4 is adapted from Munzner (2014) and lists the main visual channels with which data might be encoded. The grouping and order of the figure is meaningful. Channels are grouped according

plot grammar    ->        ggplot2 spec

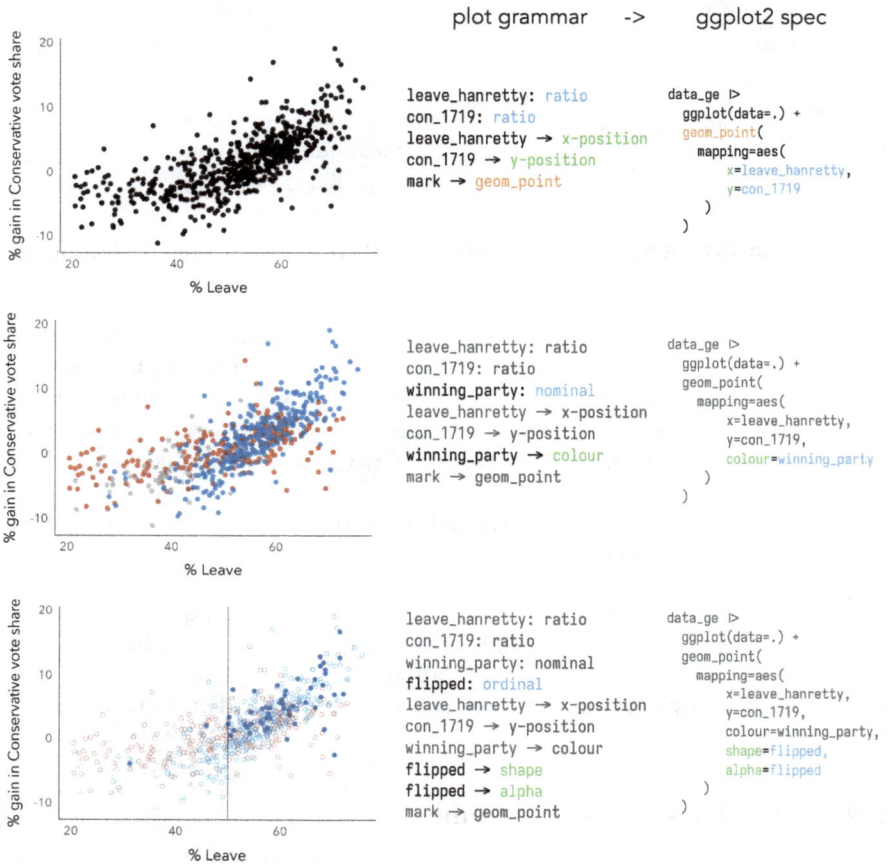

```
leave_hanretty: ratio          data_ge ▷
con_1719: ratio                  ggplot(data=.) +
leave_hanretty → x-position      geom_point(
con_1719 → y-position             mapping=aes(
mark → geom_point                     x=leave_hanretty,
                                      y=con_1719
                                  )
                                 )
```

```
leave_hanretty: ratio          data_ge ▷
con_1719: ratio                  ggplot(data=.) +
winning_party: nominal           geom_point(
leave_hanretty → x-position       mapping=aes(
con_1719 → y-position                 x=leave_hanretty,
winning_party → colour                y=con_1719,
mark → geom_point                     colour=winning_party
                                  )
                                 )
```

```
leave_hanretty: ratio          data_ge ▷
con_1719: ratio                  ggplot(data=.) +
winning_party: nominal           geom_point(
flipped: ordinal                  mapping=aes(
leave_hanretty → x-position           x=leave_hanretty,
con_1719 → y-position                 y=con_1719,
winning_party → colour                colour=winning_party,
flipped → shape                       shape=flipped,
flipped → alpha                       alpha=flipped
mark → geom_point                 )
                                 )
```

**Figure 3.3:** Plots, grammars and underlying ggplot2 specifications for the scatterplot.

to the tasks to which they are best suited and then ordered according to their effectiveness at supporting those tasks. The left grouping displays *magnitude:order* channels – those that are best suited to tasks aimed at quantifying data items. The right grouping displays *identity:category* channels – those that are most suited to supporting tasks that involve isolating and associating data items.

### 3.2.4   Evaluating designs

The effectiveness rankings of visual channels in Figure 3.4 are not simply based on Munzner's preference. They are informed by detailed experimental work by Cleveland and McGill (1984), later replicated by Heer and Bostock (2010),

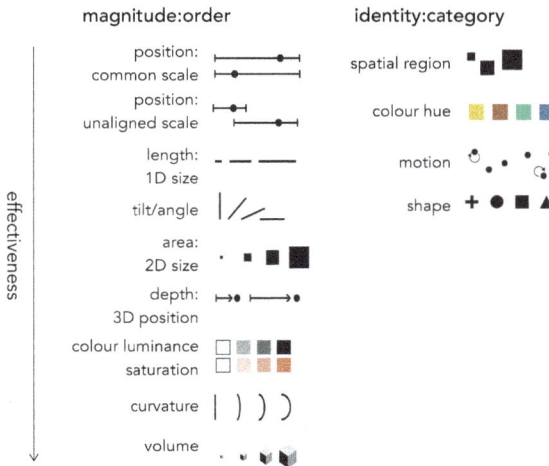

**Figure 3.4:** Visual channels to which data items can be encoded, adapted from Munzner (2014).

which involved conducting controlled experiments testing people's ability to make judgements from graphical elements. We can use Figure 3.4 to help make decisions around which data item to encode with which visual channel. This is particularly useful when designing data-rich graphics, where several data items are to be encoded simultaneously. Figure 3.4 also offers a low cost way of *evaluating* different designs against their encoding effectiveness.

To illustrate this, we can use Munzner's ranking of channels to evaluate The Washington Post graphic discussed in Figure 3.1. Table 3.2 provides a summary of the encodings used in the graphic. US counties are represented using a peak-shaped *mark*. The key purpose of the graphic is to depict the geography of voting outcomes. The most effective quantitative channel – position on an aligned scale – is used to order the county marks with a geographic arrangement. With the positional channels taken, the two quantitative measures are encoded with the next highest ranked channel, length or 1D size: height varies according to number of *total votes cast* and width according to *margin size*. The marks are additionally encoded with two categorical variables: whether the county-level result was a *landslide* and also the *winning party*. Since the intention is to give greater visual saliency to counties that resulted in a landslide, this is an ordinal variable encoded with a quantitative channel: area / 2D size. The *winning party*, a categorical nominal variable, is encoded using colour hue.

Each of the encoding choices follow conventional wisdom in that data items are encoded using visual channels appropriate to their measurement level.

**Table 3.1:** Encoding effectiveness for Gamio and Keating's (2016) Washington Post graphic that emphasises *vote margin and size* of counties using triangle marks.

| Data item | Type | Channel | Rank |
|---|---|---|---|
| Magnitude:Order | | | |
| County location | interval | position in x,y | 1. quant |
| Total votes cast | ratio | length | 3. quant |
| Margin size | ratio | length | 3. quant |
| Is landslide | ordinal | area | 5. quant |
| Identity:Category | | | |
| Winning party | nominal | colour hue | 2. cat |

Glancing down the "rank" column, the graphic has high effectiveness. While technically *spatial region* is the most effective channel for encoding nominal data, it is already in use as the marks are arranged by geographic position. Additionally, it makes sense to distinguish **Republican** and **Democrat** wins using the colours with which they are always represented. Given the fact that the positional channels represent geographic *location*, length to represent *votes cast* and *vote margin*, the only superior visual channel to 2D area that could be used to encode the *landslide* variable is *orientation*. There are very good reasons for not varying the orientation of the arrow marks. Most obvious is that this would undermine perception of length encodings used to represent the vote margin (width) and absolute vote size (height).

> **i** Visualization design and trade-offs
>
> Data visualization design almost always involves trade-offs. A general principle is to identify and prioritise data and analysis tasks, then match the most effective encodings to the data and tasks that have the greatest priority. Less important data items and tasks therefore get less effective encodings. In practice, visualization design involves exercising more creative thinking – it is sometimes preferable to defy convensional wisdom in order to provoke some desired response. Either way, good visualization design is sensitive to this interplay between tasks, data and encoding.

## 3.2.5   Symbolisation

"Symbolization is the process of encoding something with meaning in order to represent something else. Effective symbol design requires that the relationship between a symbol and the information that symbol represents (the referent) be clear and easily interpreted."

White (2017)

---

Implicit in the discussion above, and when making design decisions, is the importance of *symbolisation*. From the original Washington Post article, the overall pattern that can be discerned is of population-dense coastal and metropolitan counties voting Democrat – densely-packed, tall, wide and blue ⋀ marks – contrasted with population-sparse rural and small town areas voting Republican – short, wide and red ⋀ marks. The graphic evokes a distinctive landscape of voting behaviour, emphasised by its caption: "*The peaks and valleys of Trump and Clinton's support*".

*Symbolisation* is used equally well in a variant of the graphic emphasising two-party *Swing* between the 2012 and 2016 elections (Figure 3.5). Each county is represented as a | mark. The *Swing* variable is then encoded by continuously varying mark angles: counties swinging **Republican** are angled to the right /; counties swinging **Democrat** are angled to the left \. Although *angle* is a less effective channel at encoding quantities than is *length*, there are obvious links to the political phenomena in the symbolisation – angled right for counties that moved to the right politically. There are further useful properties in this example. Since county voting is spatially auotocorrelated, we quickly assemble from the graphic dominant patterns of Swing to the Republicans (Great Lakes, rural East Coast), predictable Republican stasis (the Midwest) and more isolated, locally exceptional swings to the Democrats (rapidly urbanising counties in the deep South).

---

**Task 1**

Complete the description table below to identify each *data item* encoded in Figure 3.5 along with its *measurement level, visual mark* and *visual channel* and the effectiveness rank of this encoding, according to Munzner (2014).

| Data item | Measurement level | Visual mark | Visual channel | Rank |
|---|---|---|---|---|
| County location | <enter here> | <enter here> | <enter here> | <enter here> |
| ... | ... | ... | ... | ... |
| ... | ... | ... | ... | ... |
| ... | ... | ... | ... | ... |
| ... | ... | ... | ... | ... |
| ... | ... | ... | ... | ... |

## HOW THE COUNTRY SWUNG TO THE RIGHT
### VAST SWATHS OF THE NATION VOTED MORE REPUBLICAN THAN IN 2012

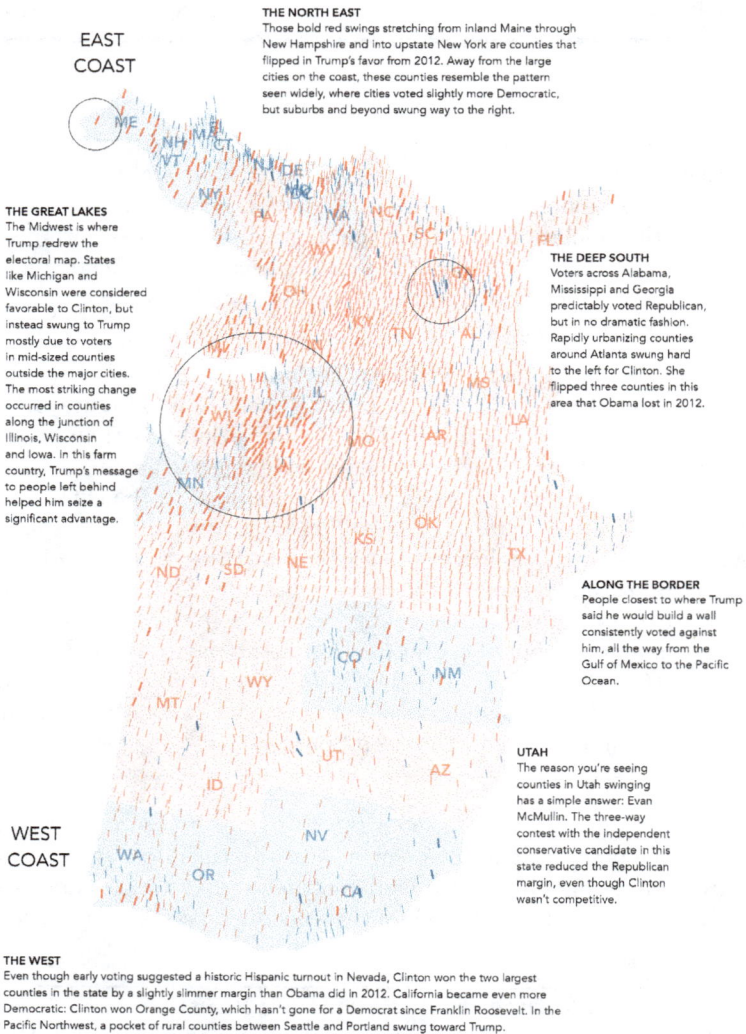

EAST
COAST

**THE NORTH EAST**
Those bold red swings stretching from inland Maine through
New Hampshire and into upstate New York are counties that
flipped in Trump's favor from 2012. Away from the large
cities on the coast, these counties resemble the pattern
seen widely, where cities voted slightly more Democratic,
but suburbs and beyond swung way to the right.

**THE GREAT LAKES**
The Midwest is where
Trump redrew the
electoral map. States
like Michigan and
Wisconsin were considered
favorable to Clinton, but
instead swung to Trump
mostly due to voters
in mid-sized counties
outside the major cities.
The most striking change
occurred in counties
along the junction of
Illinois, Wisconsin
and Iowa. In this farm
country, Trump's message
to people left behind
helped him seize a
significant advantage.

**THE DEEP SOUTH**
Voters across Alabama,
Mississippi and Georgia
predictably voted Republican,
but in no dramatic fashion.
Rapidly urbanizing counties
around Atlanta swung hard
to the left for Clinton. She
flipped three counties in this
area that Obama lost in 2012.

**ALONG THE BORDER**
People closest to where Trump
said he would build a wall
consistently voted against
him, all the way from the
Gulf of Mexico to the Pacific
Ocean.

**UTAH**
The reason you're seeing
counties in Utah swinging
has a simple answer: Evan
McMullin. The three-way
contest with the independent
conservative candidate in this
state reduced the Republican
margin, even though Clinton
wasn't competitive.

WEST
COAST

**THE WEST**
Even though early voting suggested a historic Hispanic turnout in Nevada, Clinton won the two largest
counties in the state by a slightly slimmer margin than Obama did in 2012. California became even more
Democratic: Clinton won Orange County, which hasn't gone for a Democrat since Franklin Roosevelt. In the
Pacific Northwest, a pocket of rural counties between Seattle and Portland swung toward Trump.

Each county is a line

Swing --
% change votes
from 2012 Rep-Dem

Thick stroke
county flipped
from 2012

Colour hue
winning
party     CLINTON  TRUMP

**Figure 3.5:** Map of swing in 2016 US presidential election results. Note that
for copyright reasons this is a re-implementation in ggplot2 of Gamio and
Keating's (2016) original, which appeared in The Washington Post.

### 3.2.6   Colour

Colour is a very powerful visual channel. When considering how to encode data with colour, it is helpful to consider three properties:

- *Hue*: what we generally refer to as "colour" in everyday life – red, blue, green.
- *Saturation*: how much of a colour there is.
- *Luminance/Brightness*: how dark or light a colour is.

The ultimate rule is to use these properties of colour in a way that matches the properties of the data (Figure 3.6). Categorical nominal data – data that cannot be easily ordered – should be encoded using discrete colours with no obvious order; so colour hue. Categorical ordinal data – data whose categories can be ordered – should be encoded with colours that contain an intrinsic order; saturation or brightness (colour value) allocated into perceptually-spaced gradients. Quantitative data – data that can be ordered and contain values on a continuous scale – should also be encoded with saturation or brightness, expressed on a continuous scale. As we will discover shortly, these principles are applied by default in ggplot2, along with access to perceptually valid schemes (e.g. Harrower and Brewer 2003).

| categorical-nominal | categorical-ordinal quantitative | quantitative |
|---|---|---|
| hue-based scheme | sequential colour-value | continuous colour-value |

**Figure 3.6:** Colour schemes matched to variable measurement level.

> **i** On colour
>
> There are very many considerations when using colour to support visual data analysis and communication – more than we have space for in this chapter. Lisa Charotte-Muth's (2018) *Guide to Colours in Data Visualization*[1] is an excellent outline of the decision-space.

## 3.3   Techniques

The technical component to this chapter analyses data from the 2019 UK General Election, reported at Parliamentary Constituency level. After importing and describing the dataset, we will generate data graphics that expose patterns in voting behaviour.

---

[1] https://blog.datawrapper.de/colorguide/

- Download the `03-template.qmd`[2] file for this chapter and save it to your `vis4sds` project.
- Open your `vis4sds` project in RStudio and load the template file by clicking `File > Open File ... > 03-template.qmd`.

### 3.3.1   Import

The template file lists the required packages – `tidyverse` and `sf` – and links to the 2019 UK General Election dataset, stored on the book's accompanying data repository. These data were initially collected via the `parlitools` R package, which is no longer maintained.

The data frame of 2019 UK General Election data is called `bes_2019`. This stores results data released by the House of Commons Library (Uberoi, Baker, and Cracknell 2020). We can get a quick overview with a call to `glimpse(<dataset-name>)`. `bes_2019` has 650 rows, one for each parliamentary constituency, and 118 columns. In the columns are variables reporting vote numbers and shares for the main political parties for the 2019 and 2017 General Elections, as well as names and codes (IDS) for each constituency and the local authority, region and country in which they are contained.

We will replicate some of the visual data analysis in Beecham (2020). For this we need to calculate an additional variable, Butler Swing (Butler and Van Beek 1990): the average change in share of the vote won by two parties contesting successive elections. Code for calculating this variable, named `swing_con_lab`, is in the `03-template.qmd`. The only other dataset to load is a `.geojson` file containing simplified geometries of constituencies, originally from ONS Open Geography Portal. This is a special class of data frame containing a Simple Features (Pebesma 2018) `geometry` column.

### 3.3.2   Summarise

You may be familiar with the result of the 2019 General Election, a landslide Conservative victory that confounded expectations. To start, we can quickly compute some summary statistics around the vote. In the code below, we count the number of seats won and overall vote share by party. For the vote share calculation, the code is a little more elaborate than we might wish at this stage. We need to reshape the data frame using `pivot_wider()` such that each row represents a vote for a party in a constituency. From here the vote share for each party can be easily computed.

```
# Number of constituencies won by party.
bes_2019 |>
```

---

[2]`https://vis4sds.github.io/vis4sds/files/03-template.qmd`

```
  group_by(winner_19) |>
  summarise(count=n()) |>
  arrange(desc(count))
## # A tibble: 11 x 2
##    winner_19                        count
##    <chr>                            <int>
##  1 Conservative                       365
##  2 Labour                             202
##  3 Scottish National Party             48
##  4 Liberal Democrat                    11
##  5 Democratic Unionist Party            8
##  6 Sinn Fein                            7
##  7 Plaid Cymru                          4
##  8 Social Democratic & Labour Party     2
##  9 Alliance                             1
## 10 Green                                1
## 11 Speaker                              1

# Share of vote by party.
bes_2019 |>
  # Select cols containing vote counts by party.
  select(
    constituency_name, total_vote_19,
    con_vote_19:alliance_vote_19, region
  ) |>
  # Pivot to make each row a vote for a party in a constituency.
  pivot_longer(
    cols=con_vote_19:alliance_vote_19,
    names_to="party", values_to="votes"
  ) |>
  # Use some regex to pull out party name.
  mutate(party=str_extract(party, "[^_]+")) |>
  # Summarise over parties.
  group_by(party) |>
  # Calculate vote share for each party.
  summarise(vote_share=sum(votes, na.rm=TRUE)/sum(total_vote_19)) |>
  # Arrange parties descending on vote share.
  arrange(desc(vote_share))

## # A tibble: 12 x 2
##    party   vote_share
##    <chr>        <dbl>
##  1 con          0.436
##  2 lab          0.321
```

```
##  3 ld            0.115
##  4 snp           0.0388
##  5 green         0.0270
##  6 brexit        0.0201
##  7 dup           0.00763
##  8 sf            0.00568
##  9 pc            0.00479
## 10 alliance      0.00419
## 11 sdlp          0.00371
## 12 uup           0.00291
```

While the Conservative party held 56% of constituencies in 2019 election, they won only 44% of the vote. The equivalent figures for Labour were 31% and 32% respectively. And although the Conservatives gained many more constituencies than they did in 2017 (when they won just 317, 49% of constituencies) their vote share hardly shifted between those elections – in 2017 the Conservative vote share was 43%. This fact is interesting as it may suggest some movement in where the Conservative party gained its majorities in 2019.

Below are some summary statistics computed over the newly created `swing_con_lab` variable. As the Conservative and Labour votes are negligible in Northern Ireland, it makes sense to focus on Great Britain for our analysis of Conservative-Labour Swing, and so the first step in the code is to create a new data frame filtering out Northern Ireland.

```
data_gb <- bes_2019 |>
  filter(region != "Northern Ireland") |>
  # Also recode to 0 Chorley and Buckingham, incoming/outgoing speaker.
  mutate(
    swing_con_lab=if_else(
      constituency_name %in% c("Chorley", "Buckingham"), 0,
      0.5*((con_19-con_17)-(lab_19-lab_17))
    )
  )


data_gb |>
  summarise(
    min_swing=min(swing_con_lab),
    max_swing=max(swing_con_lab),
    median_swing=median(swing_con_lab),
    num_swing=sum(swing_con_lab>0),
    num_landslide_con=sum(con_19>50, na.rm=TRUE),
    num_landslide_lab=sum(lab_19>50, na.rm=TRUE)
  )
```

```
## # A tibble: 1 x 6
## min_swing max_swing median_swing num_swing num_land_con num_land_lab
##      <dbl>     <dbl>        <dbl>     <int>        <int>        <int>
## 1    -6.47      18.4         4.44       599          280          120
```

### 3.3.3   Plot distributions

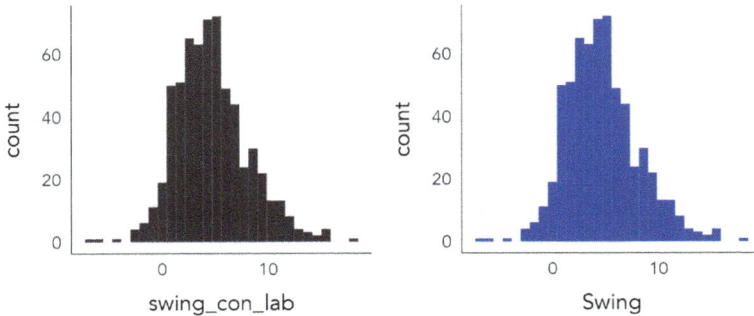

**Figure 3.7:** Histograms of Butler two-party Labour-Conservative Swing.

Let's start with ggplot2 specifications by plotting some of these variables. Below is the code for plotting a histogram of the Swing variable.

```
data_gb |>
  ggplot(mapping=aes(swing_con_lab)) +
  geom_histogram()
```

A reminder of the general form of a ggplot2 specification:

1. Start with some *data*: `data_gb`.
2. Define the *encoding*: `mapping=aes()` into which we pass the `swing_con_lab` variable.
3. Specify the *marks* to be used: `geom_histogram()` in this case.

Different from the scatterplot example, there is more happening in the internals of ggplot2 when creating a histogram. The Swing variable is partitioned into bins, and observations in each bin are counted. The x-axis (bins) and y-axis (counts by bin) are derived from the `swing_con_lab` variable.

By default the histogram's bars are given a grey colour. To *set* them to a different colour, add a `fill=` argument to `geom_histogram()`. In the code block below, colour is set using hex codes. The term *set*, not *map* or *encode*, is used for principled reasons. Any part of a ggplot2 specification that involves encoding data – mapping a data item to a visual channel – should be specified through the `mapping=aes()` argument. Anything else, for example changing the default colour, thickness and transparency of marks, needs to be *set* outside of this argument.

```
data_gb |>
  ggplot(mapping=aes(swing_con_lab)) +
  geom_histogram(fill="#003c8f") +
  labs(x="Swing", y="count")
```

You will notice that different elements of a ggplot2 specification are added (+) as layers. In the example above, the additional layer of labels (`labs()`) is not intrinsic to the graphic. However, often you will add layers that do affect the graphic itself. For example, the scaling of encoded values (e.g. `scale_*_continuous()`) or whether the graphic is to be conditioned on another variable to generate small multiples for comparison (e.g. `facet_*()`). Adding a call to `facet_*()`, we can compare how Swing varies by region (Figure 3.8). The plot is annotated with the *median* value for Swing (4.4) by adding a vertical line layer (`geom_vline()`) set with an x-intercept at this median value. From this, there is some evidence of a regional geography to the 2019 vote: London and Scotland are distinctive in containing relatively few constituencies swinging greater than the expected midpoint; North East, Yorkshire & The Humber, and to a lesser extent West and East Midlands, appear to show the largest relative number of constituencies swinging greater than the midpoint.

> Task 2
>
> Update the earlier ggplot2 specification to produce a set of histograms of the Swing variable faceted by region, similar to that in Figure 3.8.

### 3.3.4   Plot ranks/magnitudes

Previously we calculated overall vote shares by political party. We could continue the exploration of votes by region, re-using this code to generate plots displaying vote shares by region, using marks and encoding channels that are suitable for magnitudes.

To generate a bar chart similar to Figure 3.9 the ggplot2 specification would be:

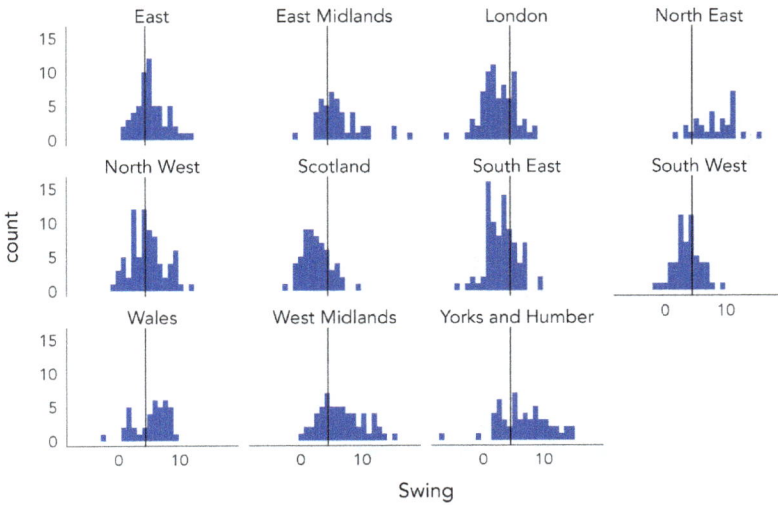

**Figure 3.8:** Histograms of Swing variable, grouped by region.

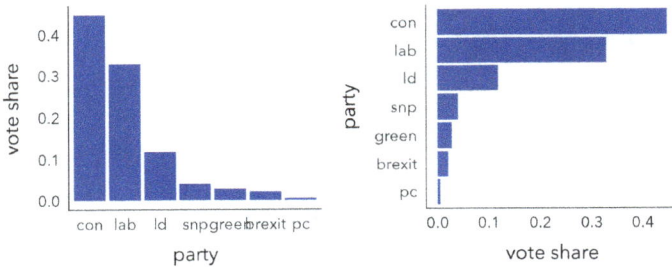

**Figure 3.9:** Plots of vote shares by party.

```
data_gb |>
  # The code block summarising vote by party.
  <some dplyr code> |>
  # Ordinal x-axis (party, reordered), Ratio y-axis (vote_share).
  ggplot(aes(x=reorder(party, -vote_share), y=vote_share)) +
  geom_col(fill="#003c8f") +
  coord_flip()
```

A quick breakdown of the specification:

1. *Data*: This is the summarised data frame in which each row is a
   political party, and the column describes the vote share recorded for
   that party.

2. *Encoding*: We have dropped the call to `mapping=`. ggplot2 always looks for `aes()`, and so we can save on code clutter. In this case we are mapping `party` to the x-axis, a categorical variable made ordinal by the fact that we reorder the axis left-to-right descending on `vote_share`. `vote_share` is mapped to the y-axis – so encoded using bar length on an aligned scale, an effective channel for conveying magnitudes.

3. *Marks*: `geom_col()` for generating the bars.

4. *Setting*: Again, we've set bar colour to manually selected dark blue. Optionally we add a `coord_flip()` layer in order to display the bars horizontally. This makes the category axis labels easier to read and also seems more appropriate for the visual "ranking" of bars.

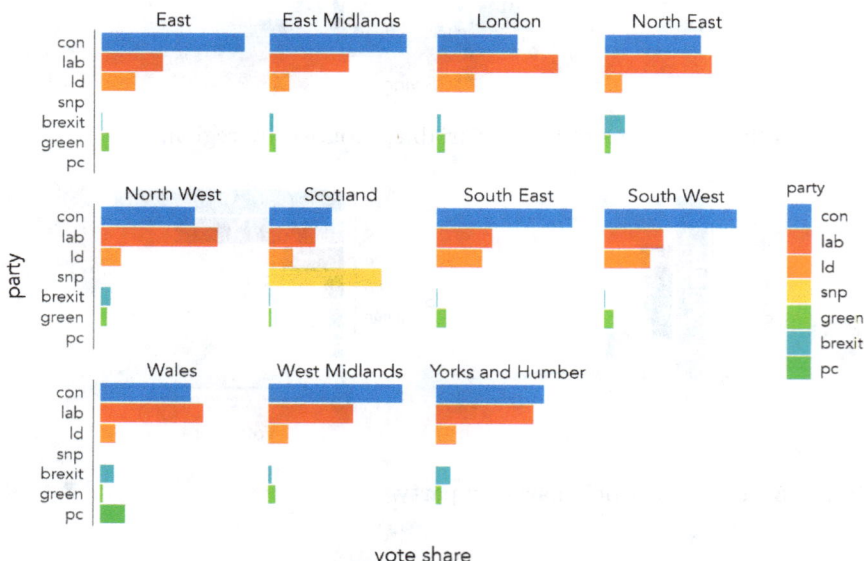

**Figure 3.10:** Plots of vote shares by party and region.

## Faceting by region

In Figure 3.10 the graphic is faceted by region. This requires an updated staged dataset grouping by `vote_share` *and* `region` and of course a faceting layer (`geom_facet(~region)`). The graphic is more data-rich, and additional cognitive effort is required in relating the political party bars between different graphical subsets. We can assist this associative task by encoding parties with an appropriate visual channel: *colour hue*. The ggplot2 specification for this is as you would expect; we add a mapping to `geom_col()` and pass the variable name `party` to the fill argument (`aes(fill=party)`).

```
data_gb |>
  # The code block summarising vote by party and also now region.
  <some dplyr code> |>
  # To be piped to ggplot2.
  ggplot(aes(x=reorder(party, vote_share), y=vote_share)) +
  geom_col(aes(fill=party)) +
  coord_flip() +
  facet_wrap(~region)
```

Trying this for yourself, you will observe that the ggplot2 internals do some thinking for us. Since `party` is a categorical variable, a categorical hue-based colour scheme is automatically applied. Try passing a quantitative variable (`fill=vote_share`) to `geom_col()` and see what happens; a quantitative colour gradient scheme is applied.

Clever as this is, when encoding political parties with colour, *symbolisation* is important. It makes sense to represent political parties using colours with which they are most commonly associated. We can override ggplot2's default colour by adding a `scale_fill_manual()` layer into which a vector of hex codes describing the colour of each political party is passed (`party_colours`). We also need to tell ggplot2 which element of `party_colours` to apply to which value of the `party` variable. In the code below, a staging table is generated summarising `vote_share` by political party and region. In the final line the `party` variable is recoded as a `factor`. You might recall from the last chapter that factors are categorical variables of fixed and orderable values – `levels`. The call to `mutate()` recodes `party` as a factor variable and orders the levels according to overall vote share.

```
# Generate staging data.
temp_party_shares_region <- data_gb |>
  select(
    constituency_name, region, total_vote_19,
    con_vote_19:alliance_vote_19
    ) |>
  pivot_longer(
    cols=con_vote_19:alliance_vote_19,
    names_to="party", values_to="votes"
    ) |>
  mutate(party=str_extract(party, "[^_]+")) |>
  group_by(party, region) |>
  summarise(vote_share=sum(votes, na.rm=TRUE)/sum(total_vote_19)) |>
  filter(
    party %in% c("con", "lab", "ld", "snp", "green", "brexit", "pc")
    ) |>
```

```
mutate(party=factor(party,
    levels=c("con", "lab", "ld", "snp", "green", "brexit", "pc"))
  )
```

Next, a vector of objects is created containing the hex codes for the colours of political parties (`party_colours`).

```
# Define colours.
con <- "#0575c9"
lab <- "#ed1e0e"
ld <- "#fe8300"
snp <- "#ebc31c"
green <- "#78c31e"
pc <- "#4e9f2f"
brexit <- "#25b6ce"
other <- "#bdbdbd"

party_colours <- c(con, lab, ld, snp, green, brexit, pc)
```

The ggplot2 specification is then updated with the `scale_fill_manual()` layer:

```
temp_party_shares_region |>
  ggplot(aes(x=reorder(party, vote_share), y=vote_share)) +
  geom_col(aes(fill=party)) +
  scale_fill_manual(values=party_colours) +
  coord_flip() +
  facet_wrap(~region)
```

---

**i Grammar of Graphics-backed visualization toolkits**

The idea behind visualization toolkits such as ggplot2 is to insert visual approaches into a data scientist's workflow. Rather than being overly concerned with low-level aspects of drawing, mapping data values to screen coordinates and scaling factors, you instead focus on aspects relevant to the analysis – the variables in a dataset and how they will be encoded and conditioned using visuals. Hadley Wickham talks about a grammar of *interactive* data analysis, whereby dplyr functions are used to rapidly prepare data for charting before being piped (`|>`) to ggplot2.

The process of searching for, defining and inserting manual colour schemes for creating Figure 3.10 might seem inimical to this. There is some reasonably involved dplyr and a little regular expression in the data preparation code that you should not be overly concerned with. Having control of these slightly more low-level properties is, though, sometimes necessary even for exploratory analysis, in this case for effecting *symbolisation* that supports comparison.

### 3.3.5 Plot relationships

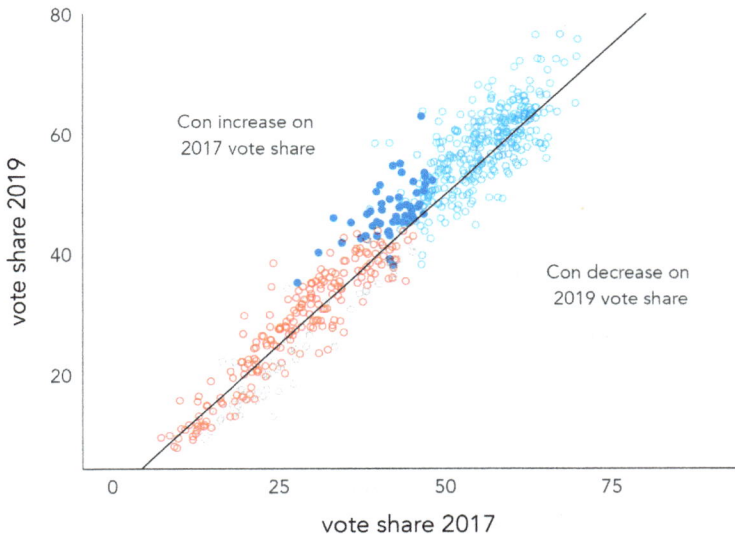

**Figure 3.11:** Plots of 2019 versus 2017 vote shares.

To continue the investigation of change in votes for the major parties between 2017 and 2019, Figure 3.11 contains a scatterplot of Conservative vote share in 2019 (y-axis) against vote share in 2017 (x-axis). The graphic is annotated with a diagonal line. If constituencies voted in 2019 in exactly the same way as 2017, the points would converge on the diagonal. Points above the diagonal indicate a larger Conservative vote share than 2017, those below the diagonal a smaller Conservative vote share than 2017. Points are coloured according to the winning party in 2019, and constituencies that flipped from Labour to Conservative are emphasised using transparency and shape.

The code for generating most of the scatterplot in Figure 3.11 is below.

```
data_gb |>
  mutate(winner_19=case_when(
         winner_19 == "Conservative" ~ "Conservative",
         winner_19 == "Labour" ~ "Labour",
```

```
         TRUE ~ "Other"
    )) |>
ggplot(aes(x=con_17, y=con_19)) +
geom_point(aes(colour=winner_19), alpha=.8) +
geom_abline(intercept = 0, slope = 1) +
scale_colour_manual(values=c(con,lab,other)) +
...
```

There is little surprising here:

1.  *Data*: The `data_gb` data frame. Values of `winner_19` that are not *Conservative* or *Labour* are recoded to *Other* using a conditional statement. This is to ease and narrow the comparison to the two major parties.
2.  *Encoding*: Conservative vote shares in 2017 and 2019 are mapped to the x- and y- axes respectively and `winner_19` to colour. `scale_colour_manual()` is used for customising the colours.
3.  *Marks*: `geom_point()` for generating the points of the scatterplot; `geom_abline()` for drawing the reference diagonal.

---

**Task 3**

The code block above doesn't exactly reproduce the graphic in Figure 3.11. Try updating the ggplot2 specification to emphasise constituencies that flipped from Labour to Conservative.

Hint: you may wish to generate a variable recording constituencies that flipped between 2017 and 2019 and encode some visual channel in the graphic on this.

---

### 3.3.6   Plot geography

The data graphics above suggest that the composition of Conservative and Labour voting may be shifting. Paying attention to the geography of voting, certainly to *changes* in voting between 2017 and 2019 elections (e.g. Figure 3.8), may therefore be instructive. We end the technical component to the chapter by generating thematic maps of the results data.

To do this we need to generate a join on the boundary dataset loaded at the start of this technical section (`cons_outline`):

```
# Join constituency boundaries.
data_gb <- cons_outline |>
```

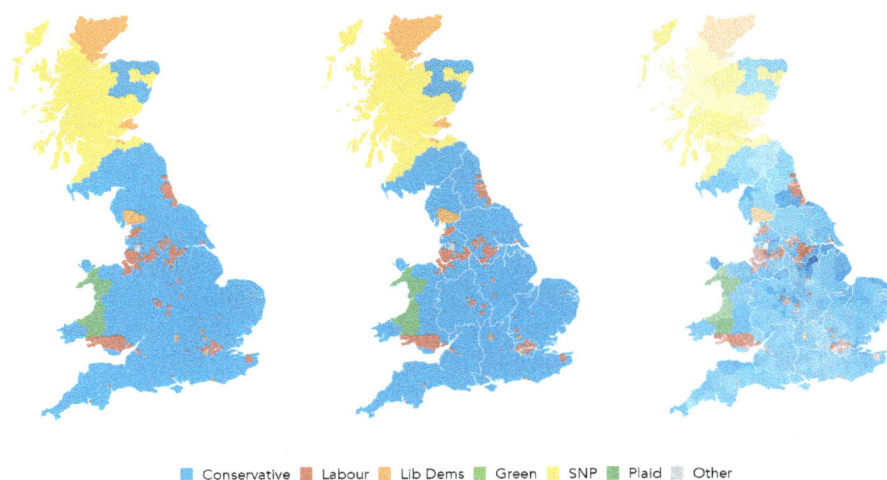

**Figure 3.12:** Choropleth of elected parties in 2019 General Election.

```
inner_join(data_gb, by=c("pcon21cd"="ons_const_id"))
# Check class.
## [1] "sf"          "data.frame"
```

The code for generating the choropleth maps of winning party by constituency in Figure 3.12:

```
# Recode winner_19 as a factor variable for assigning colours.
data_gb <- data_gb |>
  mutate(
    winner_19=if_else(winner_19=="Speaker", "Other", winner_19),
    winner_19=factor(winner_19, levels=c("Conservative", "Labour",
    "Liberal Democrat", "Scottish National Party", "Green",
    "Plaid Cymru", "Other"))
    )
party_colours <- c(con, lab, ld, snp, green, pc, other)
# Plot map.
data_gb |>
  ggplot() +
  geom_sf(aes(fill=winner_19), colour="#eeeeee", linewidth=0.01) +
  # Optionally add a layer for regional boundaries.
  geom_sf(data=. %>% group_by(region) %>% summarise(),
      colour="#eeeeee", fill="transparent", linewidth=0.08) +
```

```
coord_sf(crs=27700, datum=NA) +
scale_fill_manual(values=party_colours)
```

A breakdown of the ggplot2 spec:

1. *Data*: Update `data_gb` by recoding `winner_19` as a factor and defining a named vector of colours to supply to `scale_fill_manual()`. Note that we also use the `party_colours` object created for the region bar chart.
2. *Encoding*: No surprises here – `fill` according to `winner_19`.
3. *Marks*: `geom_sf()` is a special class of geometry. It draws objects using the contents of a simple features data frame's (Pebesma 2018) `geometry` column. In this case `MULTIPOLYGON`, so read this as a polygon shape primitive.
4. *Coordinates*: `coord_sf` – we set the coordinate system (CRS) explicitly. In this case OS British National Grid.
5. *Setting*: Constituency boundaries are subtly introduced by setting the `geom_sf()` mark to light grey (`colour="#eeeeee"`) with a thin outline (`linewidth=0.01`). On the map to the right, outlines for regions are added as another `geom_sf()` layer. Note how this is achieved in the second `geom_sf()`. The `data_gb` dataset initially passed to ggplot2 (identified by the . mark) is collapsed by region (with `group_by()` and `summarise()`) and in the background the boundaries in `geometry` are aggregated by region.

---

> **i Preparing data for plotting**
>
> A general point from the code blocks in this chapter is that proficiency in `dplyr` and `tidyr` is a necessity. Throughout the book you will find yourself needing to calculate new variables, recode variables and reorganise data frames before handing them over to ggplot2 for plotting.

---

In the third map of Figure 3.12 the transparency (`alpha`) of filled constituencies is varied according to the Swing variable. This demonstrates that constituencies swinging most dramatically for Conservative (darker colours) are in the midlands and North of England and not in London and the South East. The pattern is nevertheless a subtle one; transparency (colour luminance / saturation) is not a highly effective visual channel for encoding quantities.

It may be worth applying the same encoding to Butler two-party swing as that used in the Washington Post graphic when characterising Republican-Democrat swing in 2016 US Elections (e.g. Beecham 2020). This can be achieved by simply adding another ggplot2 layer, though the code is a little more involved. ggplot2's `geom_spoke()` primitive draws line segments parameterised by a location (x- y-position) and angle. With this we can encode constituencies with | marks that angle to the right / where the constituency swings towards Conservative and to the left where it swings towards Labour \. This encoding better exposes the pattern of constituencies forming Labour's "red wall" in the north of England, as well as parts of Wales and the Midlands flipping to Conservative.

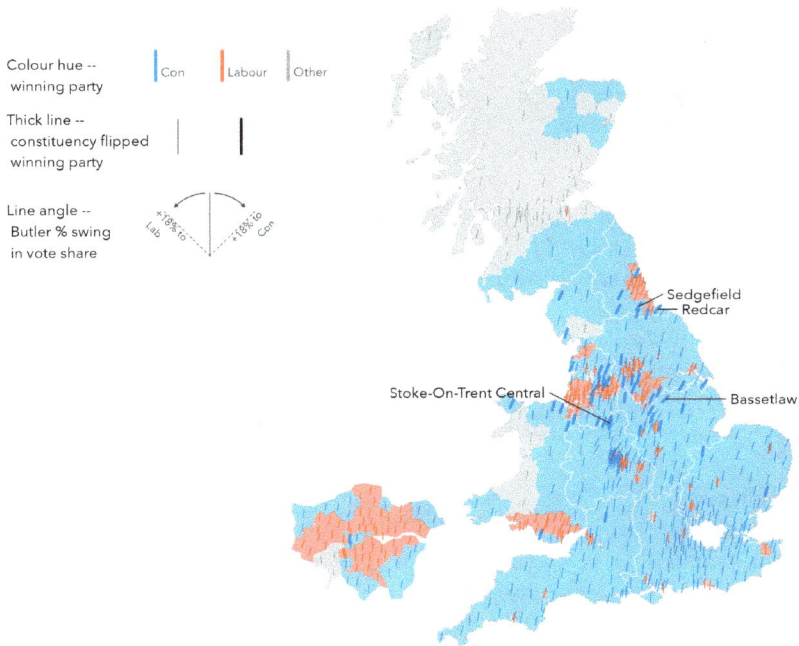

**Figure 3.13:** Map of Butler Con-Lab Swing in 2019 General Election.

The ggplot2 specification:

```
# Find the maximum Swing values to pin the min and max angles to.
max_shift <- max(abs(data_gb |> pull(swing_con_lab)))
min_shift <- -max_shift


# Re-define party_colours to contain just three values: hex codes for
# Conservative, Labour and Other.
party_colours <- c(con, lab, other)
names(party_colours) <- c("Conservative", "Labour", "Other")


# Plot Swing map.
data_gb |>
  mutate(
    is_flipped=seat_change_1719 %in%
```

```
      c("Conservative gain from Labour",
      "Labour gain from Conservative"),
  elected=if_else(!winner_19 %in% c("Conservative", "Labour"), "Other",
      as.character(winner_19)),
      swing_angle=
        get_radians(map_scale(swing_con_lab,min_shift,max_shift,135,45)
      )
  ) |>
ggplot()+
geom_sf(aes(fill=elected), colour="#636363", alpha=.2, linewidth=.01)+
geom_spoke(
  aes(x=bng_e, y=bng_n, angle=swing_angle, colour=elected,
    linewidth=is_flipped),
  radius=7000, position="center_spoke"
  )+
coord_sf(crs=27700, datum=NA)+
scale_linewidth_ordinal(range=c(.2,.5))+
scale_colour_manual(values=party_colours)+
scale_fill_manual(values=party_colours)
```

A breakdown:

1.  *Data*: `data_gb` is updated with a Boolean (TRUE/FALSE) variable iden-
    tifying whether or not the constituency flipped between successive
    elections (`is_flipped`), and a variable simplifying the party elected to
    either Conservative, Labour or Other. `swing_angle` contains the angles
    used to orient the line marks. A convenience function (`map_scale()`)
    pins the maximum swing values to 45 degrees and 135 degrees –
    respectively max swing to the right, Conservative and max swing to
    the left, Labour.

2.  *Encoding*: `geom_sf()` is again filled by elected party. This encoding
    is made more subtle by adding transparency (`alpha=.2`). `geom_spoke()`
    is mapped to the geographic centroid of each Constituency (`bng_e` -
    easting, `bng_n` - northing), coloured on elected party, sized on whether
    the constituency flipped its vote and tilted or angled according to
    the `swing_angle` variable.

3.  *Marks*: `geom_sf()` for the constituency boundaries, `geom_spoke()` for
    the angled line primitives.

4.  *Scale*: `geom_spoke()` primitives are sized to emphasise whether con-
    stituencies have flipped. The size encoding is censored to two values
    with `scale_linewidth_ordinal()`. Passed to `scale_colour_manual()` and
    `scale_fill_manual()` is the vector of `party_colours`.

5.  *Coordinates*: `coord_sf` – the CRS is OS British National Grid, so

we define constituency centroids using easting and northing planar coordinates.

6. *Setting*: The `radius` of `geom_spoke()` lines is a sensible default arrived at through trial and error, its `position` set using a newly created `center_spoke` class.

There are helper functions that must also be run to execute the ggplot2 code above correctly. In order to position lines using `geom_spoke()` centred on their x- y- location, we need to create a custom ggplot2 subclass. Details are in the `03-template.qmd` file. Again, this is somewhat involved for a chapter introducing ggplot2 for analysis. Nevertheless, hopefully you can see from the plot specification above that the principles of mapping data to visuals can be implemented straightforwardly in ggplot2: line marks for constituencies (`geom_spoke()`), positioned in x and y according to British National Grid easting and northings and oriented (`angle`) according to two-party Swing.

## Dot-density maps

A common design challenge when presenting population data spatially is to accurately reflect *geography* at the same time as the *quantitative* outcome of interest – in this case, the location and shape of constituencies versus their associated vote sizes. You may be familiar with cartogram layouts used in electoral analysis. They are maps that distort geographic space in order to size constituencies according to the voting population rather than their physical area. Dot-density maps also convey absolute numbers of votes but in a way that preserves geography. In the example below, each dot represents 1,000 votes for a given party – Conservative, Labour, Other – and dots are positioned in the constituencies from which those votes were made. Dots therefore concentrate in population-dense areas of the country.

The difficulty in generating dot-density maps is not wrangling ggplot2, but in preparing data to be plotted. We need to create a randomly located point within a constituency's boundary for every thousand votes that are made there. R packages specialised to dot-density maps provide functions for doing this, but it is quite easy to achieve using the sorts of functional and `tidyverse`-style code introduced throughout this book. We include the code for Figure 3.14 directly below the plot. While compact, there are some quite advanced functional programming concepts (in the use of `purrr::map()`) that we do not explain. These concepts are in fact covered in some detail and with proper description in later chapters of the book.

```
# Collect 2019 GE data from which dots are approximated.
vote_data <- bes_2019 |>
  filter(ons_const_id!="S14000051") |>
  mutate(
```

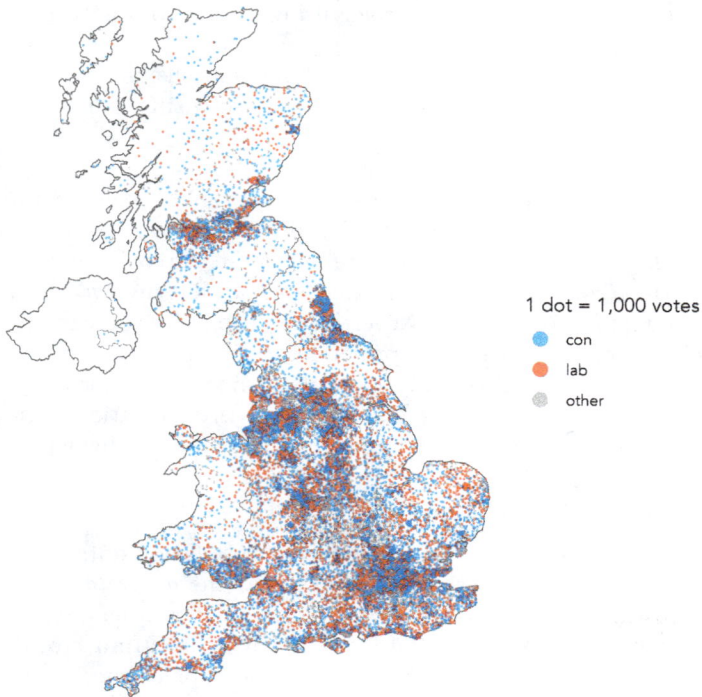

**Figure 3.14:** Dot density map of 2019 General Election result.

```
    other_vote_19=total_vote_19-(con_vote_19 + lab_vote_19)
    ) |>
select(
    ons_const_id, constituency_name, region, con_vote_19,
    lab_vote_19, other_vote_19
    ) |>
pivot_longer(
    cols=con_vote_19:other_vote_19,
    names_to="party", values_to="votes"
    ) |>
mutate(
    party=str_extract(party, "[^_]+"),
    votes_dot=round(votes/1000,0)
    ) |>
filter(!is.na(votes_dot))

# Sample within constituency polygons.
```

```
# This might take a bit of time to execute.
start_time <- Sys.time()
sampled_points <-
  cons_outline |>
  select(geometry, pcon21cd) |> filter(pcon21cd!="S14000051") |>
  inner_join(
    vote_data |> group_by(ons_const_id) |>
    summarise(votes_dot=sum(votes_dot)) |> ungroup(),
    by=c("pcon21cd"="ons_const_id")
    ) |>
  nest(data=everything()) |>
  mutate(
    sampled_points=map(data,
      ~sf::st_sample(
        x=.x, size=.x$votes_dot, exact=TRUE, type="random"
        ) |> st_coordinates() |>
        as_tibble(.name_repair=~c("east", "north"))),
      const_id=map(data, ~.x |> st_drop_geometry() |>
      select(pcon21cd, votes_dot) |> uncount(votes_dot))
    ) |>
  unnest(-data) |>
  select(-data)
end_time <- Sys.time()
end_time - start_time
point_votes <- vote_data |> select(party, votes_dot) |>
  uncount(votes_dot)
sampled_points  <- sampled_points |>  bind_cols(point_votes)

# Plot sampled points.
party_colours <- c(con, lab, other)
sampled_points |>
  ggplot() +
  geom_sf(
    data=cons_outline, fill="transparent",
    colour="#636363", linewidth=.03
    ) +
  geom_sf(data=cons_outline |>
      inner_join(vote_data, by=c("pcon21cd"="ons_const_id")) |>
      group_by(region) |>  summarise(),
    fill="transparent", colour="#636363", linewidth=.1) +
  geom_point(
    aes(x=east,y=north, fill=party, colour=party),
    alpha=.5, size=.6, stroke=0
    )+
```

```
scale_fill_manual(values=party_colours, "1 dot = 1,000 votes")+
scale_colour_manual(values=party_colours, "1 dot = 1,000 votes")+
guides(colour=guide_legend(override.aes=list(size=3)))+
theme_void()
```

## 3.4   Conclusions

Visualization design is ultimately a process of decision-making. Data must be filtered and prioritised before being encoded with marks, visual channels and symbolisation. The most successful data graphics are those that expose structure, connections and comparisons that could not be achieved easily via other, non-visual means. This chapter has introduced concepts – a vocabulary, framework and empirically-informed guidelines – that help support this decision-making and that underpin modern visualization toolkits, ggplot2 especially. Through an analysis of UK 2019 General Election data, we have demonstrated how these concepts can be applied in a real data analysis.

## 3.5   Further Reading

For a primer on visualization design principles:

- Munzner, T. 2014. "Visualization Analysis and Design", Boca Raton, FL: *CRC Press.*

A paper presenting evidence-backed guidelines on visualization design, aimed at applied researchers:

- Franconeri S. L., Padilla L. M., Shah P., Zacks J. M., Hullman J. (2021). "The science of visual data communication: What works". *Psychological Science in the Public Interest,* 22(3), 110–161. doi: 10.1177/15291006211051956.

For an introduction to ggplot2 and its relationship with Wilkinson's (1999) *grammar of graphics:*

- Wickham, H., Çetinkaya-Rundel, M., Grolemund, G. 2023, "R for Data Science, 2nd Edition", Sebastopol, CA: *O'Reilly.*
  – Chapters 2, 10.

Excellent paper looking at consumption and impact of election forecast visualizations:

- Yang, F. et al. 2024. "Swaying the Public? Impacts of Election Forecast Visualizations on Emotion, Trust, and Intention in the 2022 U.S. Midterms." *IEEE Transactions on Visualization and Computer Graphics*, 30(1), 23–33. doi: 10.1109/TVCG.2023.3327356.

# 4

## *Exploratory Data Analysis*

By the end of this chapter you should gain the following knowledge and practical skills.

---

Knowledge outcomes

- ☐ Understand how data graphics can support exploratory data analysis (EDA).
- ☐ The main chart types, their benefits and drawbacks, for analysing variation within- and between- variables.
- ☐ Three strategies for effecting comparison in EDA – juxtaposition, superposition and explicit encoding (Gleicher et al. 2011) – and the role of models in elevating comparisons by codifying expectation.
- ☐ Appreciate that EDA *is* scientific practice: knowledge is developed in light of data and models.

---

Skills outcomes

- ☐ ggplot2 code that applies encodings and layouts to support comparison.
- ☐ `tidyverse`-style code for calculating model-expected values and residuals.

---

## 4.1 Introduction

Exploratory Data Analysis (EDA) is an approach that aims to expose the properties and structure of a dataset, and from here suggest analysis directions. In an EDA relationships are quickly inferred, anomalies are labelled, models are suggested and evaluated. EDA relies heavily on visual approaches to analysis; it is common to generate many dozens of often throwaway data graphics when exploring a dataset for the first time.

This chapter demonstrates how the concepts and principles introduced previously, of data types and their visual encoding, can be applied to support EDA. It does so by analysing STATS19, a dataset containing detailed information on every reported road traffic crash in Great Britain that resulted in personal injury. STATS19 is highly detailed, with many categorical variables. The chapter starts by revisiting commonly used chart types for analysing within-variable variation and between-variable co-variation in a dataset. It then focuses more directly on the STATS19 case, and how detailed comparison across many categorical variables can be effected using colour, layout and statistical computation.

## 4.2   Concepts

### 4.2.1   Exploratory data analysis and statistical graphics

In Exploratory Data Analysis (EDA), graphical and statistical summaries are used to build knowledge and understanding of a dataset. The goal of EDA is to infer relationships, identify anomalies and test new ideas and hypotheses. Rather than a formal set of techniques, EDA should be considered a sort of outlook to analysis. It aims to reveal properties, patterns and relationships in a dataset, and from there expectations, codified via models, to be further investigated in context via more targeted data graphics and statistics.

The early stages of an EDA may be very data-driven. Datasets are described abstractly according to their measurement level and corresponding data graphics and summary statistics generated from these descriptions (e.g. Table 4.1). As knowledge and understanding of the dataset increases, researchers might apply more targeted theory and prior knowledge when developing and evaluating models.

Visual approaches play an important role in both these stages of analysis. For example, when first examining variables according to their shape and location, data graphics help identify patterns that statistical summaries miss, such as whether variables are multi-modal, the extent and direction of outliers. When more specialised models are proposed, graphical summaries can add important detail around where, and by how much, the observed data depart from the model. It is for this reason that data visualization is seen as intrinsic to EDA (John W. Tukey 1977).

**Table 4.1:** A breakdown of statistical and graphical summaries performed in EDA based on variable measurement types.

| Measurement type | Statistic | Chart type |
|---|---|---|
| Within-variable variation | | |
| Nominal | mode \| entropy | bar charts, dot plots ... |
| Ordinal | median \| percentile | bar charts, dot plots ... |
| Continuous | mean \| variance | histograms, box plots, density plots ... |
| Between-variable variation | | |
| Nominal | contingency tables | mosaic/spine plots ... |
| Ordinal | rank correlation | slope/bump charts ... |
| Continuous | correlation | scatterplots, parallel coordinate plots ... |

## 4.2.2 Plots for continuous variables

**Within-variable variation**

Figure 4.1 presents statistical graphics that are commonly used to show variation in continuous variables measured on a ratio and interval scale. In this instance, the graphics summarise the age of pedestrians injured (`casualty_age`) for a random sample of recorded road crashes.

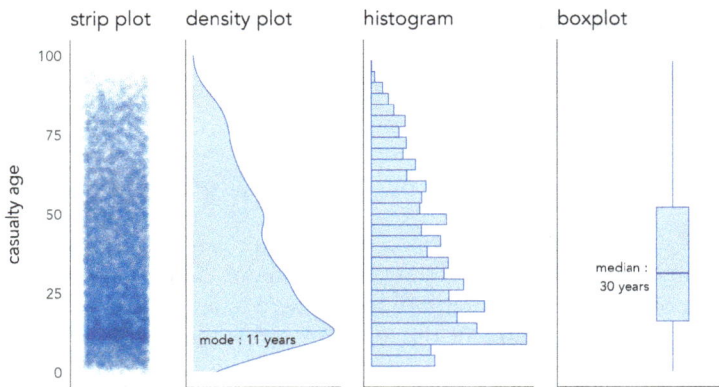

**Figure 4.1:** Univariate plots of dispersion in casualty ages from a sample of pedestrian-vehicle road crashes.

In the left-most column is a *strip-plot*. Every observation is displayed as a dot and mapped to `y-position`, with dots horizontally perturbed and made a little transparent. Although strip-plots scale poorly, the advantage is that all observations are displayed without needing to impose some aggregation. It is possible to visually identify the 'location' of the distribution – denser dots towards the youngest ages (c.20 years) – but also that there is spread across the age values.

*Histograms* partition observations into equal-range bins, and observations in each bin are counted. These counts are encoded on an aligned scale using bar length. Increasing the number of the bins increases the resolution of the graphical summary. If reasonable decisions are made around choice of bin, histograms give distributions a shape that is expressive. It is easy to identify the location of a distribution, to see that it is uni- bi- or multi-modal. Different from the strip-plot, the histogram shows that despite the heavy spread, the distribution of `casualty_age` is right-skewed, and we'd expect this given the location of the mean (36 years) relative to the median (30 years).

A problem with histograms is the potential for discontinuities and artificial edge-effects around the bins. *Density plots* overcome this and can be thought of as smoothed histograms. They show the probability density function of a variable – the relative amount of probability attached to each value of `casualty_age`. From glancing at the density plots, an overall shape to the distribution can be immediately derived. It is also possible to infer statistical properties – the mode of the distribution, the peak density, the mean and median – by a sort of visual averaging and approximating the midpoint of the area under the curve.

Finally, *boxplots* (McGill and Larsen 1978) encode these statistical properties directly. The box is the interquartile range (IQR) of the `casualty_age` variable, the vertical line splitting the box is the median, and the whiskers are placed at observations $\leq 1.5*$IQR. While we lose important information around the shape of a distribution, boxplots are space-efficient and useful for comparing many distributions at once.

---

**i** Inequalities in who-hit-whom (by age)

Since the average age of pedestrian road casualties is surprisingly low, it may be instructive to explore the distribution of `casualty_age` conditioning on another variable differentiated using colour. Figure 4.2 displays boxplots and density plots of the location and spread in `casualty_age` by vehicle and individual for all crashes involving pedestrians. A noteworthy pattern is that riders of bicycles and motorcycles tend to be younger than the pedestrians they collide with, whereas for buses, taxis, HGVs and cars the reverse is true.

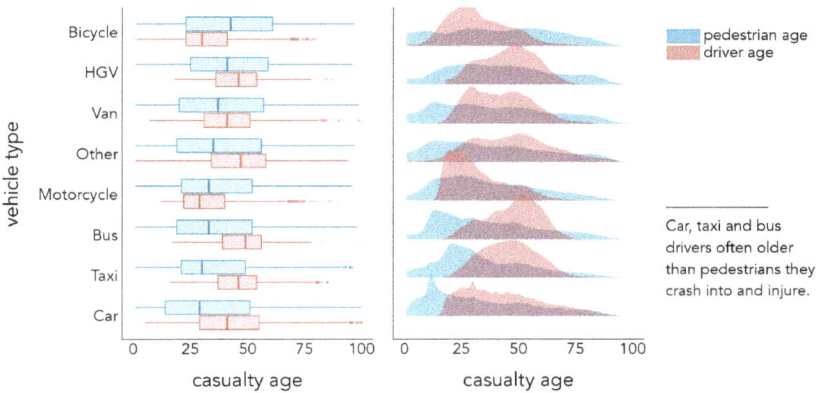

**Figure 4.2:** Boxplots of casualty age by vehicle and individual type (driver or pedestrian).

## Between-variable variation

The previous chapter included several scatterplots for exploring associations in electoral voting behaviour. Scatterplots are used to check whether the association between variables is linear, but also to make inferences about the direction and intensity of linear correlation between variables – the extent to which values in one variable depend on the values of another – and the nature of variation between variables – the extent to which variation in one variable depends on another. In an EDA it is common to quickly compare associations between many quantitive variables in a dataset using scatterplot matrices or, less often, parallel coordinate plots. There are few variables in the STATS19 dataset measured on a continuous scale, but in Chapter 6 we will use both scatterplot matrices and parallel coordinate plots when building models that attempt to structure and explain between-variable covariation, again on an electoral voting dataset.

### 4.2.3 Plots for categorical variables

#### Within-variable variation

For categorical variables, within-variable variation is judged on how relative frequencies distribute across the variable's categories. Bar charts are commonly used, as bar length is effective at encoding frequency. When analysing variation across several categories, it is useful to flip bar charts on their side so that category labels can be easily scanned and, unless there is a natural ordering, arrange the bars in descending order based on their frequency. This is demonstrated in Figure 4.3, which shows the frequencies with which different vehicles types are involved in pedestrian casualties.

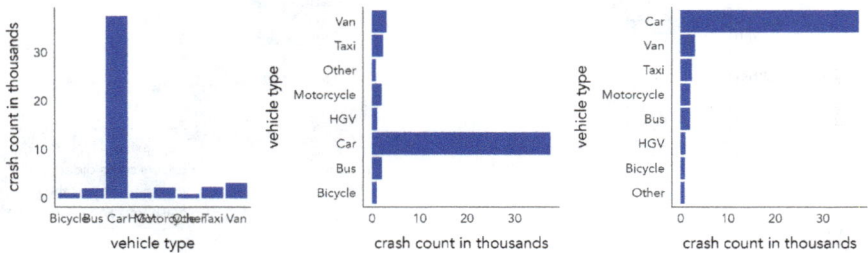

**Figure 4.3:** Bars displaying crash frequencies by vehicle type.

For summarising frequencies across many categories alternative chart types that minimise non-data-ink (Tufte 1983), such as dot plots, may be appropriate. The left plot in Figure 4.4 displays pedestrian crash counts for boroughs in London, ordered by crash frequency, grouped by whether boroughs are in inner- or outer- London and coloured on whether crashes took place on weekdays or weekends. Lines connecting dots emphasise the differences in absolute numbers between time periods. Although a consistent pattern is of greater crash counts during weekdays, the gap is less obvious for outer London boroughs; there may be relatively more pedestrian crashes occurring in central London boroughs during weekdays. The second graphic is a heatmap with the same ordering and grouping of boroughs, but with columns coloured according to crash frequencies by vehicle type, further grouped by weekday and weekend times. Remembering Munzner's (2014) ordering of visual channels, we trade-off some precision in estimation when encoding frequencies in heatmaps. A greater difficulty, irrespective of encoding channel, comes from the dominance of cars and weekdays; variation between vehicle types and time periods outside of this is almost imperceptible.

**Between-variable covariation: standardised bars and mosaic plots**

In Figure 4.4 we began to make between-category comparison; we asked whether there are relatively more or fewer crashes by time period or vehicle type in certain boroughs than others. There are chart types that explicitly support these sorts of analysis tasks. Figure 4.5 compares pedestrian crash frequencies in London by vehicle type involved and whether the crash occurred on weekdays or weekends.

First, stacked bars are ordered by frequency, distinguishing time period using colour lightness. Cars are by far the dominant travel mode, contributing the largest number of crashes resulting in injury to pedestrians. Whether or not pedestrian injuries involving cars occur more on weekends than other modes

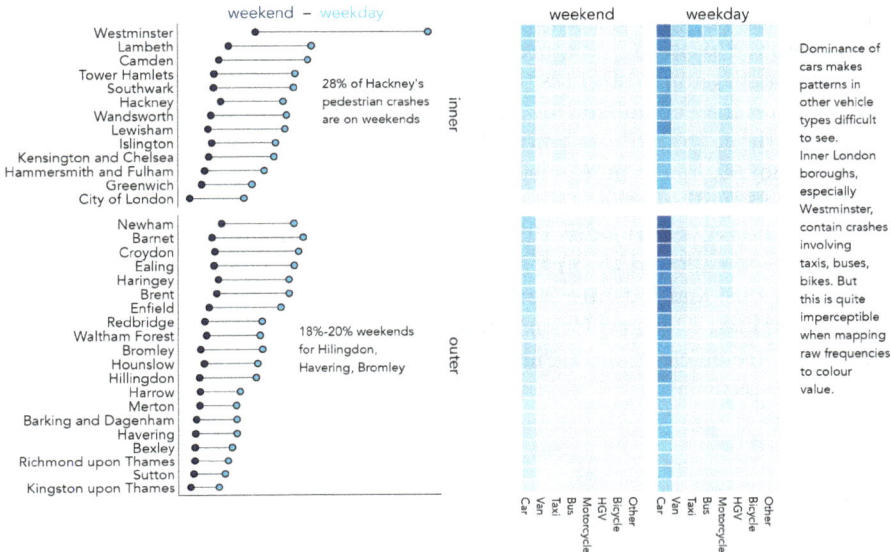

**Figure 4.4:** Cleveland dot plots and heatmaps summarising crash frequencies by London borough, period of day and vehicle type.

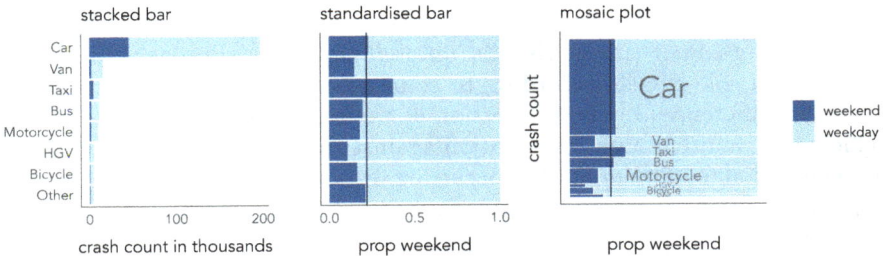

**Figure 4.5:** Bars and mosaic plot displaying association between vehicle type and injury severity.

is not clear from the left-most chart. Length encodes absolute crash counts effectively but relative comparison on time periods between vehicle types is challenging. In standardised bars the absolute length of bars is fixed, and bars are split according to proportional weekday / weekend crashes (middle). The plot is also annotated according to the *expected proportion* of weekday crashes if crashes occurred by time period independently of vehicle type (22%). This encoding shows pedestrian crashes involving taxis occur more than would be expected at weekends, while the reverse is true of crashes involving vans, bikes and HGVs. However, we lose a sense of the absolute numbers involved.

Failing to encode absolute number – the amount of information in support of some observed pattern – is a problem in EDA. Since proportional summaries are agnostic to sample size, they can induce false discoveries, overemphasising patterns that may be unlikely to replicate in out-of-sample tests. It is sometimes desirable, then, to update standardised bar charts so that they are weighted by frequency: to make more visually salient those categories that occur more often and visually downweight those that occur less often. This is possible using mosaic plots (Friendly 1992; Jeppson and Hofmann 2023). Bar widths and heights are allowed to vary, so bar area is proportional to absolute number of observations, and bars are further subdivided for relative comparison across category values. Mosaic plots are useful tools for exploratory analysis. That they are space-efficient and regularly sized also means they can be flexibly laid out for comparison.

> **i** ggmosaic
>
> The mosaic plot in Figure 4.5 was generated using the ggmosaic package, an extension to ggplot2 (Jeppson and Hofmann 2023).

**Encoding variation from expectation**

The heatmap in Figure 4.4 is hampered by the dominating effect of cars and weekdays. Any additional structure by time period and vehicle type in boroughs outside of this is visually unintelligible. An alternative approach could be to colour cells according to some relevant effect size statistic: for example, differences in the proportion of weekend crashes occurring in any vehicle-type and borough combination against the global average proportion, or expectation, of 22% of crashes occurring on weekends. A problem with this approach is that at this more disaggregated level, sample sizes become quite small. Large proportional differences could be encoded that are nevertheless based on negligible differences in absolute crash frequencies.

There are measures of effect size sensitive both to absolute and relative differences from expectation. Signed chi-score residuals (Visvalingam 1981), for example, represent expected values as *counts* separately for each category combination in a dataset – in this case, pedestrian crashes recorded in a borough in a stated time period involving a stated vehicle type. Observed counts $(O_i...O_n)$ are then compared to expected counts $(E_i...E_n)$ as below:

$$\chi = \frac{O_i - E_i}{\sqrt{E_i}}$$

The way in which differences between observed and expected values (residuals) are standardised in the denominator is important. If the denominator was simply the raw expected value, the residuals would express the proportional difference between each observation and its expected count value. The denominator is instead transformed using the square root ($\sqrt{E_i}$), which has the effect of inflating smaller expected values and squashing larger expected values, thereby adding saliency to differences from expectation that are also large in absolute number.

Figure 4.6 updates the heatmaps with signed residuals encoded using a diverging colour scheme (Brewer and Campbell 1998) – red for cells with greater crash counts than expected, blue for cells with fewer crash counts than expected. The assumption in the first heatmap is that crash counts by borough distribute independently of vehicle type. Laying out the heatmap such that inner and outer London boroughs are grouped for comparison is instructive: fewer than expected crashes in inner London are recorded for cars; greater than expected for all other vehicle types but especially taxis and bicycles. This pattern is strongest (largest residuals) for very central boroughs, where pedestrian crash frequencies are also likely to be highest and where cars are comparatively less dominant as a travel mode. For almost all boroughs, again especially central London boroughs, there is a clear pattern of modes other than cars, taxis and buses overrepresented amongst crashes occurring on weekdays, again reflecting the transport dynamics of the city.

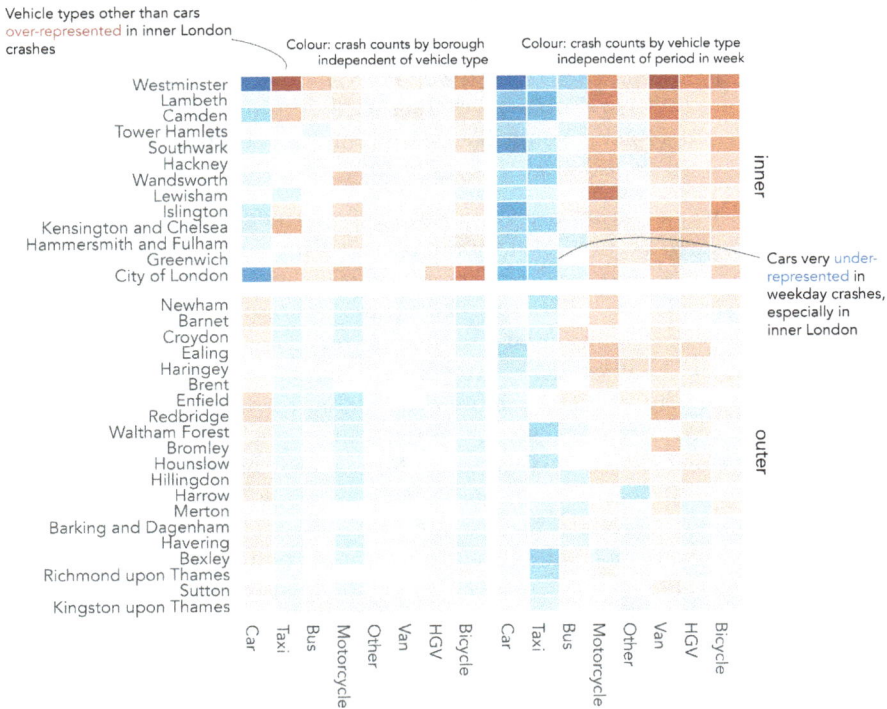

**Figure 4.6:** Heatmaps of crashes by vehicle type and period of week for London Boroughs.

**Table 4.2:** Implementing Gleicher et al.'s (2011) three comparison strategies in ggplot2.

| Strategy | Function | Use |
|---|---|---|
| Juxtaposition | `faceting` | Create separate plots in rows and/or columns by conditioning on a categorical variable. Each plot has same encoding and coordinate space. |
| Juxtaposition | `patchwork` `cowplot pkg` | Flexibly arrange plots of different data type, encoding and coordinate space. |
| Superposition | `geoms` | Layering marks on top of each other. Marks may be of different data types but must share the same coordinate space. |
| Explicit encoding | `NA` | No strategy specialised to explicit encoding. Often variables cross 0, so diverging schemes, or graphics with clearly annotated and symbolised thresholds are used. |

### 4.2.4   Strategies for supporting comparison

A key role for data graphics in EDA is in supporting comparison. Three strategies typically deployed are *juxtaposition, superposition* and *explicit encoding* (see Gleicher et al. 2011). Table 4.2 defines each and identifies how they can be implemented in ggplot2. You will see these implementations being deployed in different ways as the book progresses.

As with most visualization design, each involves trade-offs , and so careful decision-making. In Figure 4.4 dotplots representing counts of weekend and weekday crashes are *superposed* on the same coordinate space, with connecting lines added to emphasise difference. This strategy is fine where two categories of similar orders of magnitude are compared, but if instead all eight vehicle types were to be encoded with categories differentiated using colour hue, the plot would be substantially more challenging to process. In Figure 4.6, comparison by vehicle type is instead effected using explicit encoding – residuals coloured above or below an expectation. Notice also the use of *containment, juxtaposition* and *layout* in both plots. By containing frequencies for inner- and outer- London boroughs in separate juxtaposed plots, and within each laying out cells top-to-bottom on frequency, the systematic differences in the composition of vehicle types involved in crashes between inner- and outer-London can be inferred.

### Comparison and layout

Layout is an extremely important mechanism for enabling comparison. The heatmaps in Figure 4.6, for example, would be much less effective were some default alphabetical ordering of boroughs used. This applies especially when exploring geography. Spatial relations are highly complex and notoriously difficult to model. It would be hard to imagine how the sorts of comparisons in the Washington Post graphics in the previous chapter (Figure 3.1 and Figure 3.5) could be made without using graphical methods, laying out the peak and line marks with a geographic arrangement in this case.

Figure 4.7 borrows the earlier mosaic plot design to study crash frequencies (bar height) and relative number of weekday/weekend crashes (dark bar width), with frequencies compared between London boroughs. Rather than laying out boroughs top-to-bottom on frequency, boroughs are given an approximate spatial arrangement. This is generated using the `gridmappr` (Beecham 2024) R package, which we describe and explore properly in Chapter 5. This arrangement enables several patterns to be quickly inferred. We can observe that pedestrian crashes involving motorcycles generally occur more in central London boroughs; those involving cars occur in greater relative number during weekends, especially so for those in central London boroughs; and, different from other vehicle types, pedestrian crashes involving cars occur in similarly large numbers in outer London boroughs (Barnet, Croydon) as they do inner London boroughs such as Westminster.

## 4.3   Techniques

The technical element to this chapter continues with the STATS19 dataset. Rather than a how-to guide for generating exploratory analysis plots in R, the section aims to demonstrate a workflow for exploratory visual data analysis:

1.  Expose pattern(s)
2.  Model expectation(s) derived from pattern(s)
3.  Show deviation from expectation(s)

It does so by exploring the characteristics of individuals involved in pedestrian crashes, with a special focus on inequalities. Research suggests those living in more deprived neighbourhoods are at elevated risk of road crash injury than those living in less-deprived areas (Tortosa et al. 2021). A follow-up question is around the characteristics of those involved in crashes: To what extent do drivers share demographic characteristics with the pedestrians they crash into, and does this vary by the location in which crashes take place?

### 4.3.1   Import

*   Download the `04-template.qmd`[1] file for this chapter, and save it to your `vis4sds` project.

---

[1]`https://vis4sds.github.io/vis4sds/files/04-template.qmd`

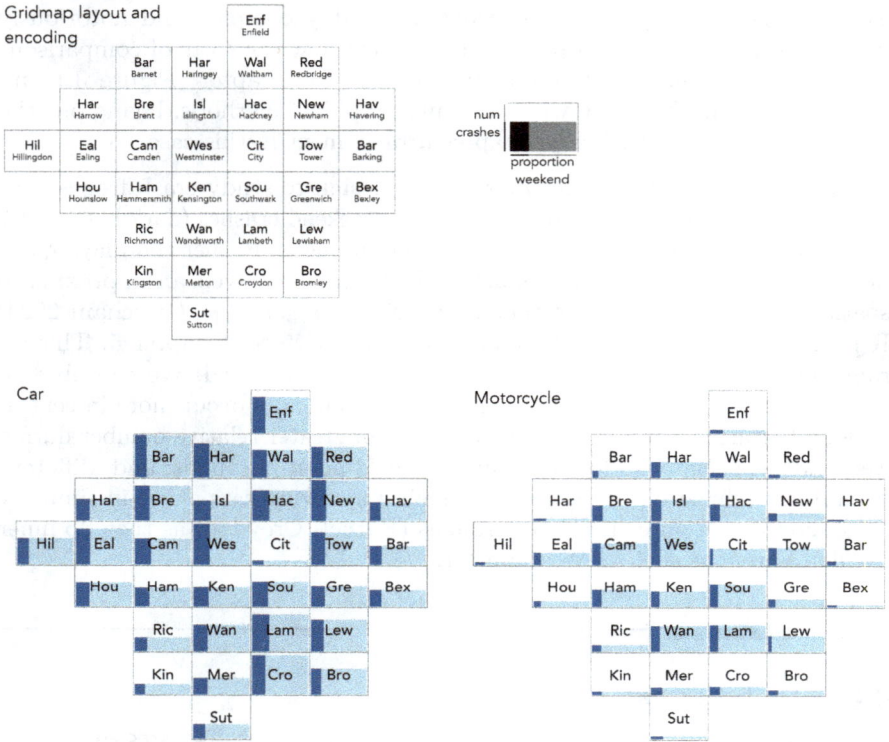

**Figure 4.7:** Mosaic plots of vehicle type and period for London Boroughs with an approximate spatial arrangement.

- Open your vis4sds project in RStudio, and load the template file by clicking File > Open File ... > 04-template.qmd.

The presented analysis is based on that published in Beecham and Lovelace (2023) and investigates vehicle–pedestrian crashes in STATS19 between 2010 and 2019, where the Index of Multiple Deprivation (IMD) class of the pedestrian, driver and crash location is recorded. Raw STATS19 data are released by the Department for Transport, but can be accessed via the stats19 R package (Lovelace et al. 2019). The data are organised into three tables:

- *Accidents* (or *Crashes*): Each observation is a recorded road crash with a unique identifier (accident_index), date (date), time (time) and location (longitude, latitude). Many other characteristics associated with the crashes are also stored in this table.
- *Casualties*: Each observation is a recorded casualty that resulted from a road crash. The *Crashes* and *Casualties* data can be linked via the accident_index

variable. As well as `casualty_severity` (`Slight`, `Serious`, `Fatal`), information on casualty demographics and other characteristics is stored in this table.

- *Vehicles*: Each observation is a vehicle involved in a crash. Again *Vehicles* can be linked with *Crashes* and *Casualties* via the `accident_index` variable. As well as the vehicle type and manoeuvre being made, information on driver characteristics is recorded in this table.

An `.fst` dataset that uses these three tables to record pedestrian crashes with associated casualty and driver characteristics has been stored in the book's accompanying data repository[2].

## 4.3.2 Sample

The focus of our analysis is inequalities in the characteristics of those involved in pedestrian crashes. There is only high-level information on these characteristics in the STATS19 dataset. However, the Index of Multiple Deprivation (Noble et al. 2019) quintile of the small area neighbourhood in which casualties and drivers live is recorded, and we have separately derived the IMD quintile of the neighbourhood in which crashes took place.

Not all recorded crashes contain this information, and we first create a new dataset – `ped_veh_complete` – identifying those linked crashes where the full IMD data are recorded:

```
# Complete demographics for pedestrians, drivers and crash locations.
ped_veh_complete <- ped_veh |>
  filter(
    !is.na(crash_quintile),
    !is.na(casualty_quintile),
    casualty_quintile != "Data missing or out of range",
    driver_quintile != "Data missing or out of range"
  )
```

The dataset contains c. 52,600 observations, 23% of linked pedestrian crashes. Although this is a large number, there may be some systematic bias in the types of pedestrian crashes for which full demographic data are recorded. For brevity we will not extensively investigate this bias, but below 'record completeness rates' are calculated for selected crash types. As anticipated, lower completeness rates appear for crashes coded as *Slight* in injury severity, but there are also lower completeness rates for crashes occurring in the highest deprivation quintile.

This difference in record completeness may reflect genuine differences in recording behaviour for crashes occurring in high deprivation neighbourhoods, or

---

[2]`https://github.com/vis4sds/data`

it might be a function of some confounding context. For example, one might expect crashes more serious in nature to be reported in greater detail and so have higher completeness rates. If crashes resulting in slight injuries are overrepresented in high deprivation areas, this might explain the pattern of completeness rates by deprivation area. To explore this further, in Figure 4.8 completeness rates are calculated separately by crash injury severity. This demonstrates, as expected, higher completeness rates for crashes resulting in more serious injury, but that record completeness is still lower for crashes taking place in the high deprivation neighbourhoods.

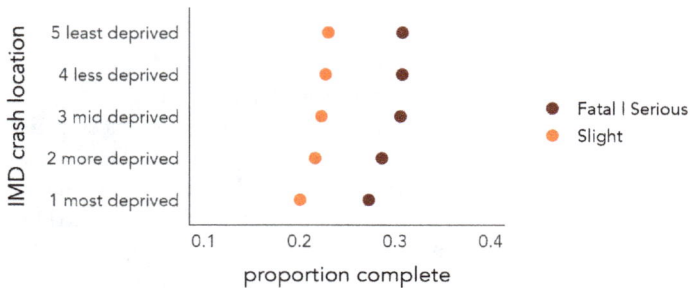

**Figure 4.8:** Completeness rates by IMD class of crash location and crash injury severity.

The code for Figure 4.8:

```
ped_veh |>
  mutate(
    is_complete=accident_index %in% (ped_veh_complete |> pull(accident_index)),
    is_ksi=if_else(accident_severity != "Slight", "Fatal | Serious", "Slight")
  ) |>
  group_by(crash_quintile, is_ksi) |>
  summarise(prop_complete=mean(is_complete)) |>
  ggplot() +
  geom_point(
    aes(y=crash_quintile, x=prop_complete, colour=is_ksi), size=2
  ) +
  scale_colour_manual(values=c("#67000d", "#fb6a4a")) +
  scale_x_continuous(limits=c(0.1,.4))
```

A breakdown of the `ggplot2` code:

1. *Data*: A Boolean variable, `is_complete`, is defined by checking `accident_index` against those contained in the `ped_veh_complete` dataset. Note that the `pull()` function extracts `accident_index` from the complete dataset as a vector of values. A Boolean variable (`is_ksi`) groups

and separates *Fatal* and *Serious* injury outcomes from those that are *Slight*. We then group on `crash_quintile` and `is_ksi` to calculate completeness rates by severity and crash location. Since `is_complete` is a Boolean value (`false=0`, `true=1`), its mean is the proportion of `true` records, in this case those with a complete status.

2. *Encoding*: Arrange dots vertically (y position) on `crash_quintile` and horizontally (x position) on `prop_complete` and `colour` on injury severity (`is_ksi`).

3. *Marks*: `geom_point()` for drawing points.

4. *Scale*: Passed to `scale_colour_manual()` are hex values for dark and light red, according to ordinal injury severity.

### 4.3.3 Abstract and relate

Now that we've identified the data for our analysis, we can begin to explore the dataset, bearing in mind the ultimate "who-hit-whom" question: Do drivers share demographic characteristics with the pedestrians they crash into, and does this vary by the location in which crashes take place?

To start, we abstract over the relevant variables: five IMD classes from high-to-low deprivation (IMD quintile 1–5) for pedestrians, drivers, and crash locations. Figure 4.9 summarises frequencies across these categories. Pedestrian crashes occur more frequently in higher deprivation neighbourhoods; those injured more often live in higher deprivation neighbourhoods; and the same applies to drivers involved in crashes. This high-level pattern is consistent with existing research (Tortosa et al. 2021) and can be explained. High deprivation areas are located in greater number in urban areas, and so we would expect greater numbers of pedestrian crashes to occur in such areas. The shapes of the bars nevertheless suggest that there are inequalities in the characteristics of pedestrians and drivers involved in crashes: frequencies are most skewed towards high deprivation bars for pedestrians and are slightly more uniform across deprivation classes for drivers. This may indicate an importing effect of drivers living in lower deprivation areas crashing into pedestrians living in higher deprivation areas – a speculative finding worth exploring.

The code for Figure 4.9:

```
ped_veh_complete |>
  select(crash_quintile, casualty_quintile, driver_quintile) |>
  pivot_longer(
    cols=everything(), names_to="location_type", values_to="imd"
    ) |>
  group_by(location_type, imd) |>
  summarise(count=n()) |> ungroup() |>
  separate(col=location_type, into="type", sep="_", extra = "drop") |>
```

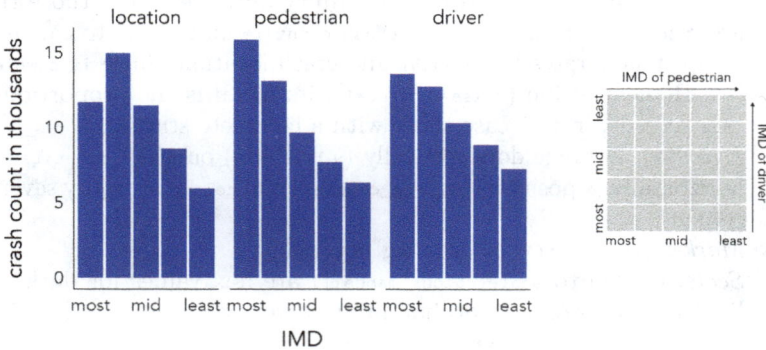

**Figure 4.9:** Frequencies of pedestrian crashes by IMD class of crash location, pedestrian injured and driver involved.

```
mutate(
  type=case_when(
    type=="casualty" ~ "pedestrian",
    type=="crash" ~ "location",
    TRUE ~ type),
  type=factor(type, levels=c("location", "pedestrian", "driver"))
) |>
ggplot() +
geom_col(aes(x=imd, y=count), fill="#003c8f") +
scale_x_discrete(labels=c("most","", "mid", "", "least")) +
facet_wrap(~type)
```

The `ggplot2` code:

1. *Data*: Select the three variables recording IMD class of crash location (`crash_quintile`), pedestrian (`casualty_quintile`) and driver (`driver_quintile`). `pivot_longer()` makes each row a crash record and IMD class; this dataset is then grouped in order to count frequencies of location, pedestrian and drivers by IMD class involved. `mutate()` is used to recode the `type` variable with more expressive labels for locations, pedestrians and drivers and to convert it to a factor in order to control the order in which variables appear in the faceted plot.

2. *Encoding*: Bars positioned vertically (`y` position) on frequency and horizontally (`x` position) on `imd` class.

3. *Marks*: `geom_col()` for drawing bars.

4. *Facets*: `facet_wrap()` for faceting on the `type` variable (location, pedestrian or driver).

### 4.3.4 Model and residual: Pass 1

To investigate how the characteristics of pedestrians and drivers *co*-vary, we can compute the joint frequency of each permutation of driver-pedestrian IMD quintile group. This results in 5x5 combinations, as in the right of Figure 4.9, and in Figure 4.10 these combinations are represented in a heatmap. Cells of the heatmap are ordered left-to-right on the IMD class of pedestrian and bottom-to-top on the IMD class of driver. Arranging cells in this way encourages linearity in the association to be emphasised. The darker blues in the diagonals demonstrate that an association between pedestrian-driver IMD characteristics exists: drivers and passengers living in similar types of neighbourhoods are involved in crashes with one another with greater frequency than those living in different types of neighbourhoods.

A consequence of the heavy concentration of crash counts, and thus colour, in the high-deprivation cells is that it is difficult to gauge variation and the strength of association in the lower deprivation cells. We can use exploratory models to support our analysis. In this case, our (unsophisticated) expectation is that crash frequencies distribute independently of the IMD class of the pedestrian-driver involved. We compute signed chi-scores describing how different the observed number of crashes in each cell position is from this expectation.

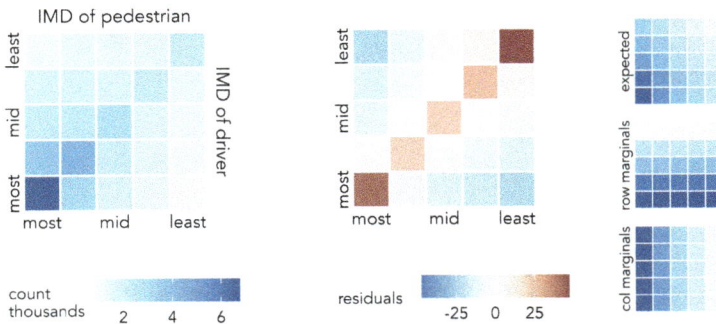

**Figure 4.10:** Pedestrian casualties by IMD quintile of pedestrian and driver.

The observed-versus-expected plot highlights the largest positive residuals are in the diagonals and the largest negative residuals are those furthest from the diagonals: we see higher crash frequencies between drivers and pedestrians living in the same or similar IMD quintiles and fewer between those in different quintiles than would be expected given no association between pedestrian-driver IMD characteristics. That the bottom left cell – high-deprivation-driver +

high-deprivation-pedestrian – is dark red can be understood when remembering that signed chi-scores emphasise effect sizes that are large in absolute as well as relative number. Not only is there an association between the characteristics of drivers and casualties, but larger crash counts are recorded in locations containing the highest deprivation and so residuals here are large. The largest positive residuals are nevertheless recorded in the top right of the heatmap – and this is more surprising. Against an expectation of no association between the IMD characteristics of drivers and pedestrians, there is a particularly high number of crashes between drivers and pedestrians living in neighbourhoods containing the lowest deprivation. An alternative phrasing: the IMD characteristics of those involved in pedestrian crashes are most narrow between drivers and pedestrians who live in the lowest deprivation quintiles.

The code:

```
model_data <- ped_veh_complete |>
  mutate(grand_total=n()) |>
  group_by(driver_quintile) |>
  mutate(row_total=n()) |> ungroup() |>
  group_by(casualty_quintile) |>
  mutate(col_total=n()) |> ungroup() |>
  group_by(casualty_quintile, driver_quintile) |>
  summarise(
    observed=n(),
    row_total=first(row_total),
    col_total=first(col_total),
    grand_total=first(grand_total),
    expected=(row_total*col_total)/grand_total,
    resid=(observed-expected)/sqrt(expected),
  )


max_resid <- max(abs(model_data$resid))


model_data |>
  ggplot(aes(x=casualty_quintile, y=driver_quintile)) +
  geom_tile(aes(fill=resid), colour="#707070", size=.2) +
  scale_fill_distiller(
    palette="RdBu", direction=-1,
    limits=c(-max_resid, max_resid)
    ) +
  coord_equal()
```

The ggplot2 spec for Figure 4.10:

1. *Data*: We create a staged dataset for plotting. Observed values

for each cell of the heatmap are computed, along with row and column marginals for deriving expected values. We assume that crash frequencies distribute independently of IMD class, and so calculate expected values for each cell of the heatmap ($E_{ij}$) from its corresponding column ($C_i$), row ($R_j$) maginals and the grand total of crashes ($GT$): $E_i = \frac{C_i \times R_i}{GT}$. The graphics in the right margin of Figure 4.10 show how expectation is spread in this way. You will notice that `group_by()` does some heavy lifting to arrive at these row, column and cell-level totals. The way in which the signed chi-score residuals are calculated in the final `group_by()` follows that described earlier in the chapter.

2. *Encoding*: Cells of the heatmap are arranged in `x` and `y` on the IMD class of pedestrians and drivers and filled according to signed chi-score residuals. .
3. *Marks*: `geom_tile()` for drawing cells of the heatmap.
4. *Scale*: `scale_fill_distiller()` for continuous ColorBrewer (Harrower and Brewer 2003) diverging scheme, using the `RdBu` palette. To make the scheme centred on 0, the maximum absolute residual value in `model_data` is used.

---

**Task 1**

You will see that data preparation with `dplyr` plays an important role in constructing data graphics in ggplot2. While the code to create `model_data` may seem somewhat cumbersome, you will find yourself reusing these data processing templates.

Now that you have seen how observed, expected and residual values can be derived for cells of a heatmap, recreate the heatmap in the left column of Figure 4.6 displaying differences from expectation in the number of pedestrian crashes by London borough, assuming that crashes distribute by borough independently of vehicle type.

---

## 4.3.5 Model and residual: Pass 2

An obvious confounding factor, neglected in the analysis above, is the IMD class of the *location* in which crashes occur. To explore this, we can condition (or facet) on the IMD class of crash location, laying out the faceted plot left-to-right on the ordered IMD classes. Eyeballing this graphic of observed counts (Figure 4.11), we see again the association between IMD characteristics for crashes occurring in the least deprived quintile and elsewhere slightly more 'mixing'. Few pedestrians living outside the most deprived quintile are involved in crashes that occur in that quintile. Given the dominating pattern is of crashes occurring in the most deprived quintiles, however, it is difficult to see too much variation from the diagonal cell in the less-deprived quintiles. An easy adjustment would be to apply a local colour scale for each faceted plot and compare relative 'leakage' from the diagonal for each IMD crash location. The more interesting question, however, is whether this known association between pedestrian and driver characteristics is stronger for certain driver-pedestrian-location combinations than others: that is, net of the dominant pattern in the top row of Figure 4.11, in which cells are there greater or fewer crash counts?

The concept that we are exploring is whether crash counts vary depending on how different the IMD characteristics of pedestrians and drivers are from those of the locations in which crashes occur. We calculate a new variable measuring this distance: 'geodemographic distance', the Euclidean distance between the IMD class of the driver-pedestrian-crash location, treating IMD as a continuous variable ranging from 1 to 5. The second row of Figure 4.11 encodes this directly. We then specify a Poisson regression model, modelling crash counts in each driver-pedestrian-crash location cell as a function of geodemographic distance for that cell. Since the range of the crash count varies systematically location-type, the model is extended with a group-level intercept that varies on the IMD class of the crash location. If regression modelling frameworks are new to you, don't worry about the specifics. More important is our interpretation and analysis of the residuals. These residuals are expressed in the same way as in the signed-chi-score model and show whether there are greater or fewer crash counts in any pedestrian-driver-location combination than would be expected given the amount of geodemographic difference between individuals and locations involved. Our expectation is that crash counts vary inversely with geodemographic distance. In EDA, we are not overly concerned with confirming this to be the case; instead we use our data graphics to explore where in the distribution, and by how much, the observed data depart from this expectation.

The vertical block of red in the left column of the left-most matrix (crashes occurring in high-deprivation areas) indicates higher than expected crash counts for pedestrians living and being hit in the most deprived quintile, both by drivers living in that high-deprivation quintile and the less-deprived quintiles (especially the lowest quintiles). This pattern is important as it persists even after having modelled for 'geodemographic distance'. There is much to unpick elsewhere in the graphic. Like many health issues, pedestrian road injuries have a heavy socio-economic element, and our analysis has identified several patterns worthy of further investigation. However, this model-backed exploratory analysis provides direct evidence of the previously suggested "importing effect" of drivers from low-deprivation areas crashing and injuring predestrians in high-deprivation areas.

Observed

Geodemographic distance

Obs vs Exp

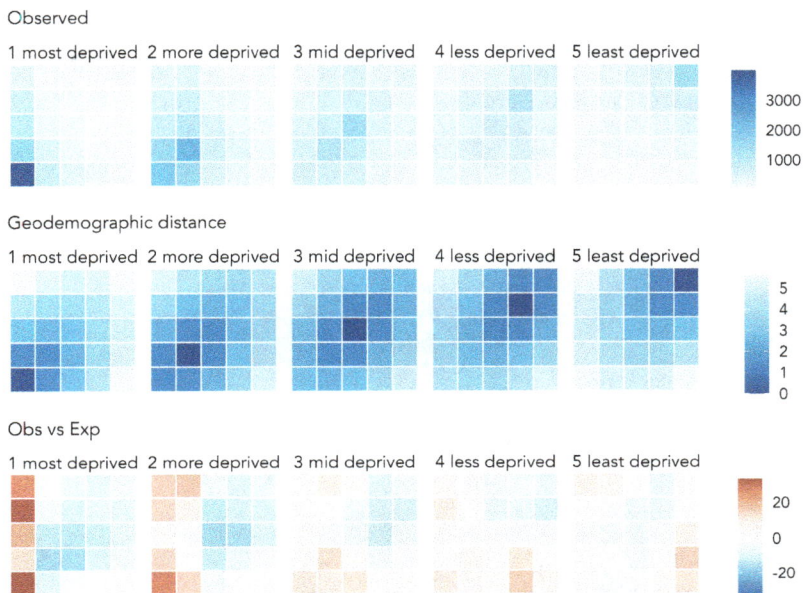

**Figure 4.11:** Pedestrian casualties by IMD quintile of pedestrian, driver and crash location.

The modelling is somewhat involved – a more gentle introduction to model-based visual analysis appears in Chapters 6 and 7 – but code for generating the model and graphics in Figure 4.11 is below.

```
model_data <- ped_veh_complete |>
  mutate(
    # Derive numeric values from IMD classes (ordered factor variable).
    across(
      .cols=c(casualty_quintile, driver_quintile, crash_quintile),
      .fns=list(num=~as.numeric(
        factor(., levels=c("1 most deprived", "2 more deprived",
        "3 mid deprived", "4 less deprived", "5 least deprived"))
        ))
    ),
    # Calculate demog_distance.
    demog_dist=sqrt(
      (casualty_quintile_num-driver_quintile_num)^2 +
      (casualty_quintile_num-crash_quintile_num)^2 +
      (driver_quintile_num-crash_quintile_num)^2
    )
```

```
) |>
# Calculate on observed cells: each ped-driver IMD class combination.
group_by(casualty_quintile, driver_quintile, crash_quintile) |>
summarise(crash_count=n(), demog_dist=first(demog_dist)) |> ungroup()


# Model crash count against demographic distance allowing the intercept
# to vary on crash quintile, due to large differences in obs frequences.
# between location quintiles.
model <- lme4::glmer(crash_count ~ demog_dist + ( 1 | crash_quintile),
                     data=model_data, family=poisson, nAGQ = 100)


# Extract model residuals.
model_data <- model_data %>%
  mutate(ml_resids=residuals(model, type="pearson"))


# Plot.
model_data |>
  ggplot(aes(x=casualty_quintile, y=driver_quintile)) +
  geom_tile(aes(fill=ml_resids), colour="#707070", size=.2) +
  scale_fill_distiller(palette="RdBu", direction=-1,
  limits=c(
    -max(abs(model_data$ml_resids)),
    max(abs(model_data$ml_resids))
  )) +
  facet_wrap(~crash_quintile, nrow=1) +
  coord_equal()
```

## Task 2: Design challenge

Key to developing data graphics in any exploratory anaysis is proficiency in implementing strategies for comparison (e.g. Table 4.2 ). Figure 4.12 makes use of juxtaposition (top graphic) and superposition + explicit encoding (bottom graphic) to compare pedestrian casualties taking place in daylight versus darkness and against the age of pedestrians injured and the IMD class of the crash location.

As expected from the analysis at the start of this chapter, there is a peak in younger adults being injured in pedestrian road crashes. For crashes taking place in darkness, this peak is less extreme and shifts to slightly 'older' young adult ages.

In total, 71% of recorded pedestrian crashes occur in daylight. The bottom graphic uses this proportion to explicitly encode an *expected* number of daylight crashes at each age group – e.g. if one were to randomly select an age and deprivation grouping, we would expect 71% of crashes to occur in daylight. This addition helps expose that there are slightly more crashes recorded in darkness for crashes taking place in mid-deprivation locations and involving younger and middle-age pedestrians. A more subtle pattern is of slightly fewer crashes in darkness than expected in older adults.

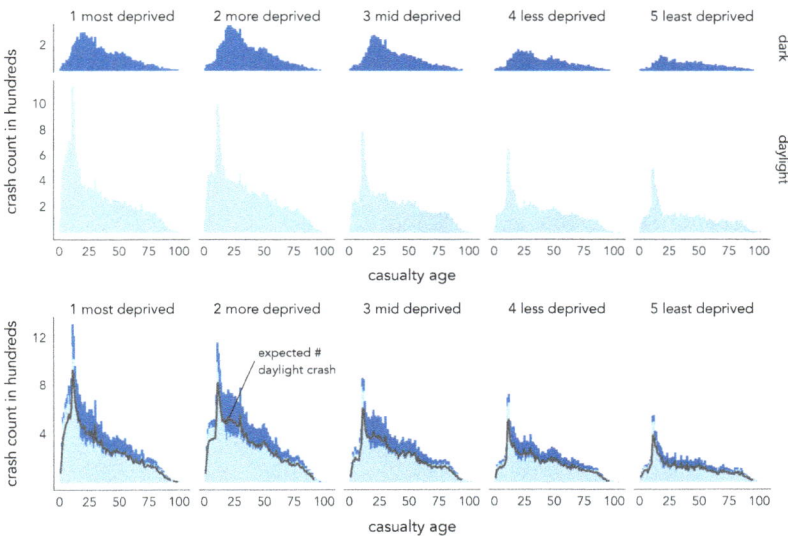

**Figure 4.12:** Pedestrian casualties by age of pedestrian, IMD quintile of crash location and level of darkness.

Try writing some `dplyr` and ggplot2 code to generate data graphics similar to Figure 4.12, or perhaps using the same layout and plot grammar but conditioning on some other category of interest in the `ped_veh` dataset. You will want to consider how the variables (`age_of_casualty`, `crash_quintile`, `light_conditions`) are mapped to visual channels via position, colour and spatial region.

## 4.4 Conclusions

Exploratory data analysis (EDA) is an approach to analysis that aims to amplify knowledge and understanding of a dataset. The idea is to explore structure, and data-driven hypotheses, by quickly generating many often throwaway statistical and graphical summaries. In this chapter we discussed chart types for exposing distributions and relationships in a dataset, depending on data type. We also showed that EDA is not model-free. Data graphics help us to see dominant patterns and from here formulate expectations that are to be modelled. Different from so-called confirmatory data analysis, however, in an EDA the goal of model-building is not to "identify whether the model fits or not [...] but rather to understand in what ways the fitted model departs from the data" (Gelman 2004). We covered visualization approaches to supporting comparison between data and expectation using juxtaposition, superimposition and explicit encoding (Gleicher et al. 2011). The chapter did not provide an exhaustive survey of EDA approaches, and certainly not an exhaustive set of chart types and model formulations for exposing distributions and relationships. By linking the chapter closely to the STATS19 dataset, we learnt a workflow for EDA that is common to most effective data analysis and communication activity:

1. Expose pattern
2. Model an expectation derived from that pattern
3. Show deviation from expectation

## 4.5 Further Reading

For discussion of exploratory analysis and visual methods in modern data analysis:

- Hullman, J. and Gelman, A. 2021. "Designing for Interactive Exploratory Data Analysis Requires Theories of Graphical Inference" *Harvard Data Science Review*, 3(3). doi: 10.1162/99608f92.3ab8a587.

Further discussion with implemented examples for road safety analysis:

- Beecham, R., and R. Lovelace. 2023. "A Framework for Inserting Visually-Supported Inferences into Geographical Analysis Workflow: Application to Road Safety Research" *Geographical Analysis*, 55: 344–366. doi: 10.1111/gean.12338.

For an introduction to exploratory data analysis in the tidyverse:

- Wickham, H., Çetinkaya-Rundel, M., Grolemund, G. 2023, "R for Data Science, 2nd Edition", Sebastopol, CA: *O'Reilly*.
  − Chapter 10.
- Ismay, C. and Kim, A. 2020. "Statistical Inference via Data Science: A ModernDive into R and the Tidyverse", New York, NY: *CRC Press*. doi: 10.1201/9780367409913.

# 5

## Geographic Networks

By the end of this chapter you should gain the following knowledge and practical skills.

> **i** Knowledge
>
> ☐ Understand the special structure and vocabulary used to represent network data.
> ☐ Appreciate the strengths, weaknesses and trade-offs of network visualizations.
> ☐ Learn design approaches for incorporating geographic context into network visualization.

> **i** Practical skills
>
> ☐ Generate semi-spatial gridmap layouts using the `gridmappr` package.
> ☐ Write `ggplot2` specifications to generate gridmaps: geographically arranged bar charts and full origin-destination maps (OD maps).
> ☐ Write code to generate model-expected values to emphasise different structure and patterns in OD maps.

## 5.1 Introduction

Networks are a special class of data used to represent things, entities, and how they relate to one another. Network data consist of two types of element: *nodes*, the entities themselves, and *edges*, the connections between nodes. Both nodes and edges can have additional information attached to them – counts, categories and directions. Network data are cumbersome to work with in R as they are not represented well by flat data frames. A common workflow is to split the data across two tables – one representing nodes and one representing edges (Wickham, Navarro, and Lin Pedersen 2023).

A category of network data used heavily in geospatial analysis is origin-destination (OD) data describing, for example, flows of bikes (Beecham et al. 2023) and commuters (Beecham and Slingsby 2019) around a city. These data consist of *nodes*, origin and destination locations, and *edges*, flows between origins and destinations. While statistics from Network Science can and have been deployed in the analysis of geospatial OD data (Y. Yang et al. 2022), visualization techniques provide much assistance in exposing the types of complex structural patterns and relations inherent in geographic flow data.

In this chapter we will work with an accessible and widely used OD network dataset: Census travel-to-work data recording counts of individuals commuting between Census geographies of the UK based on their home and workplace. Specifically, we will work with data in London recording travel-to-work between the city's 33 boroughs.

## 5.2    Concepts

### 5.2.1    Node summary

The *nodes* in this dataset are London's 33 boroughs, and the *edges* are directed OD pairs between boroughs. In Figure 5.1 frequencies of the number of jobs available in each borough and workers living in each borough (the nodes) are represented. Note that job-rich boroughs in central London – Westminster, City of London – contain many more jobs than workers residing in them. We can infer that there is a high level of in-commuting to those boroughs and the reverse, a high level of out-commuting, for worker-rich boroughs containing larger numbers of workers relative to jobs.

### 5.2.2    Node-link representations

The most common class of network visualization used to represent network data are node-link diagrams. These depict graphs in two dimensions as a force-directed layout. Nodes are positioned such that those sharing greater connection – edges with greater frequencies – are closer than those that are less well-connected – that do not share edges with such large frequencies. Edges are drawn as lines connecting nodes, and so *node-link* diagrams.

The left graphic in Figure 5.2 uses a force-directed layout to represent the travel-to-work data. Nodes, London boroughs, are sized according to the number of jobs and workers they contain and edges, commuters between boroughs, are represented as lines sized by commuter frequency. As is often the case with node-link diagrams, the graphic looks complex. Job-rich boroughs, Westminster and City of London, are labelled and have many connecting

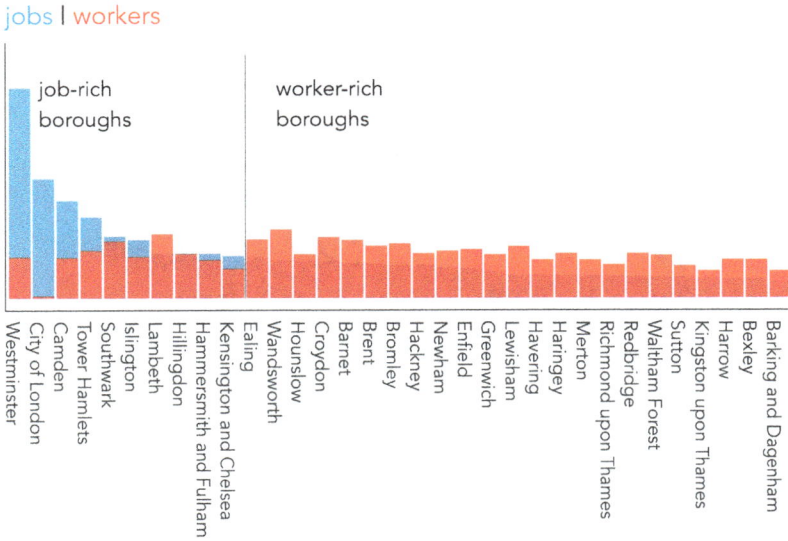

**Figure 5.1:** Barchart of jobs and workers contained in London boroughs.

lines – most likely workers commuting in from other London boroughs. Other more 'residential' boroughs are labelled. Lambeth and Wandsworth, for example, contain many connecting lines – likely residents commuting out to other London boroughs for work. That these boroughs are close in geographic space as well as force-directed space suggests that between-borough commuting is spatially dependent.

To investigate this more directly, it makes sense to *position* nodes (boroughs) with a geographic arrangement. In the right of Figure 5.2, boroughs are placed in their exact geographic position (geometric centroid of boroughs) and line width and colour are used to encode edge (commuter flow) frequency. Boroughs are again represented with circles sized according to frequency (the total number of jobs and workers contained in the borough), and commuter flow *direction* is encoded by making lines asymmetric, following Wood, Slingsby, and Dykes (2011): the straight ends are origins, the curved ends destinations.

The geographic positioning of boroughs adds context, and the encoding of direction provides further detail. For example, the pattern of commuting into central London boroughs versus more peripheral boroughs, with asymmetric commuter flows into Westminster, is just about detectable, and a more symmetric pattern between outer London boroughs is somewhat easier to see. However, there are problems that affect the usefulness of the graphic. Self-contained flows – where individuals live and work in the same borough – are not shown. The graphic is cluttered with a 'hairball' effect due to

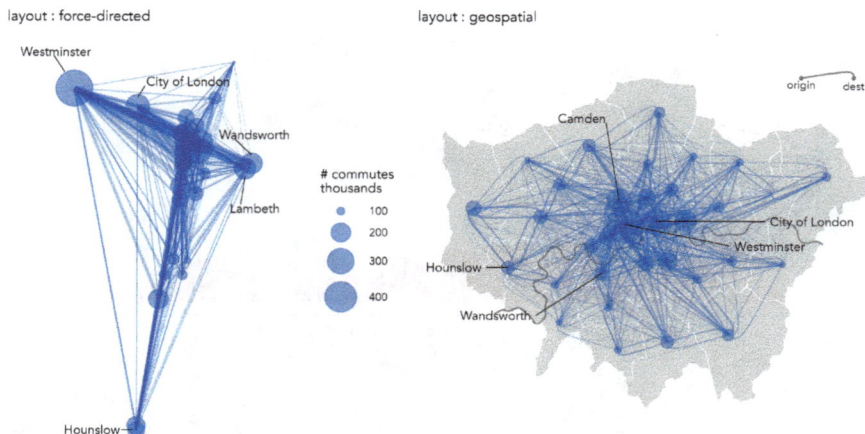

**Figure 5.2:** Flowlines with edges showing frequencies between London boroughs.

multiple overlapping lines. Longer flows appear more visually dominant than do shorter flows, an unhelpful artefact of the encoding. Also, aggregating to the somewhat arbitrary geometric centre of boroughs and drawing lines between these locations implies an undue level of spatial precision; the pattern of commuting would likely look different were individual flows encoded with precise OD locations of home and workplace.

### 5.2.3   Origin-Destination matrices

An alternative way to represent commuter flow frequencies is as an origin-destination matrix, as in Figure 5.4. The columns are destinations, London boroughs into which residents commute for work; the rows are origins, London boroughs from which residents commute out for work. Commute frequencies are encoded using colour value – the darker the colour, the larger the number of commutes between those boroughs. Boroughs are ordered left-to-right and top-to-bottom according to the total number of jobs accessed in each borough.

While using colour lightness rather than line width to show flow magnitude is a less effective encoding channel (following Munzner 2014), there are obvious advantages. The salience bias of longer flows is removed – every OD pair, 1039 in total ($33^2$), is given equal graphic saliency. Ordering cells of the matrix by destination size (number of jobs accessed in each borough) helps to emphasise patterns in the job-rich boroughs, but also encourages within and between borough comparison. For example, the lack of colour outside of the diagonals in the less job-rich boroughs, which also tend to be in outer London, suggests that labour markets there might be more self-contained. By applying a local scaling on destination (right plot), we can explore commutes into individual boroughs in a more detailed way. The vertical strips of blue for other job-rich central and inner London boroughs (Hammersmith & Fulham and Kensington & Chelsea), suggesting reasonably high-levels of in-commuting to access jobs there.

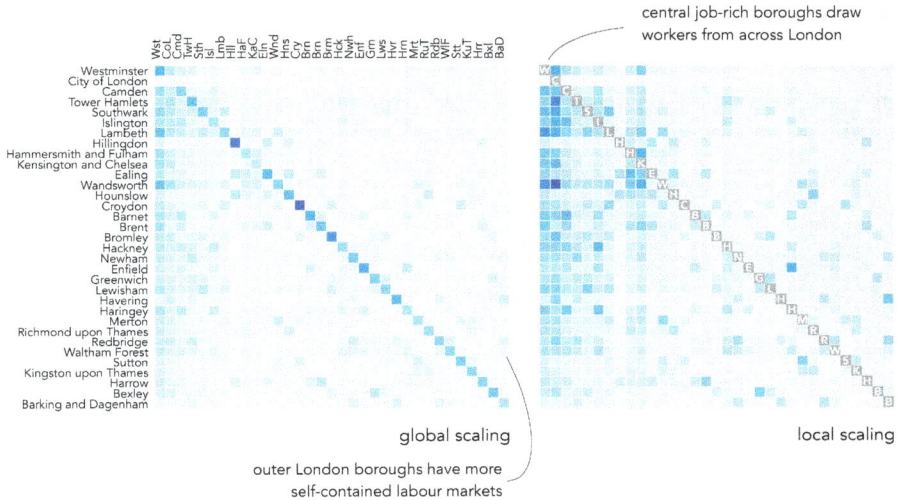

**Figure 5.3:** Origin-destination matrices ordered according to borough size on number of jobs. In the right graphic a separate 'local' colour scale is created for each destination borough.

## 5.2.4 Origin-Destination maps

The OD matrices expose new structure that could not be so easily inferred from the node-link visualizations. For phenomena such as commuting, however, the fact that geographic context is missing is a pitfall. OD maps (Wood, Dykes, and Slingsby 2010) are a form of matrix that make better use of layout and position to support this spatial dimension of analysis. They take a little to get your head around, but the idea is elegant.

OD maps contains exactly the same cells as an OD matrix, but the cells are re-ordered with an approximate geographic arrangement, as in the right column of Figure 5.4. So, for example, we may be interested in focussing on *destination*, or workplace, boroughs. In the first highlighted example, commutes into Westminster are considered (the left-most column of the OD matrix). Cells in the highlighted column are coloured according to the number of workers resident in each borough that travel into Westminster for work. In the map to the right, these cells are then re-ordered with an approximate spatial arrangement. The geographic ordering allows us to see that residents access jobs in Westminster in large numbers from many boroughs in London, but

especially from Wandsworth (Wns), Lambeth (Lmb) and Southwark (Sth) to
the south of Westminster (Wst).

In the second example – the middle row of the matrix – we focus on *origins*:
specifically, commutes out of Hackney. Cells in the highlighted row are coloured
according to the number of jobs accessed in each borough by residents living in
Hackney, but travelling out of that borough for work. Cells are again reordered
in the inset map. This demonstrates that commuting patterns are reasonably
localised. The modal destination/workplace borough remains Westminster, but
relatively large numbers of jobs are accessed in Camden (Cmd), Islington (Isl),
Tower Hamlets (TwH) and the City of London (CoL) by residents living in
Hackney.

Commutes into Westminster

Commutes out of Hackney

**Figure 5.4:** Origin-destination matrices: highlighted destination (Westminster)
and origin (Hackney) with geospatial arrangement.

OD maps extend this idea by displaying *all cells* of the OD matrix with
a geographic arrangement. This is achieved via a 'map-within-map' layout
(Figure 5.5), made possible by the fact that the gridded arrangement contains
regularly-sized cells.

Figure 5.6 is a destination-focussed OD map (D-OD). Each larger reference cell
identifies destinations, and the smaller cells are coloured according to origins
– the number of residents in each borough commuting into the reference cell
for work. The map uses a local colour scaling, with same origin-destination
cells greyed out. Flow counts are summarised over each reference borough
(destination in this case) and normalised according to the maximum flow count
for that reference borough.

The local scaling allows us to characterise the geography of commuting into
boroughs in some detail. The two job-rich boroughs, Westminster and City of

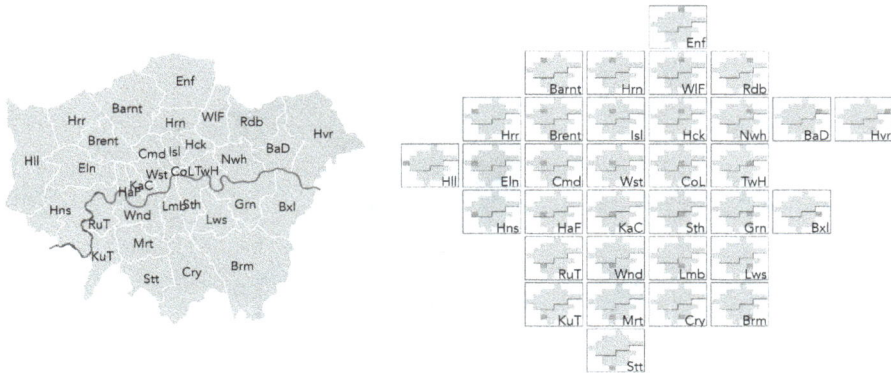

**Figure 5.5:** Map-witin-map layout required for OD maps. The approximate spatial arrangement is created by the `gridmappr` package (Beecham 2024).

London, clearly draw workers in large proportions across London boroughs, and to a lesser extent this is the case for other central/inner boroughs such as Islington (Isl), Camden (Cmd) and Tower Hamlets (TwH). For outer London boroughs, commuting patterns are more localised. Large numbers of available jobs are filled by workers living in neighbouring boroughs. Readers familiar with London's geography may notice that inner London boroughs south of the river – Lambeth (Lam), Wandsworth (Wnd), Southwark (Sth) – tend to draw workers in greater number from boroughs that are also south of the river.

---

**Task 1**

Although OD maps overcome several problems of flow-line based visualizations and share several of the characteristics of effective data graphics discussed in Chapter 3, they do require some interpretation, especially when seen for the first time.

Test your knowledge by studying Figure 5.6 and completing the following look-up tasks:

- For jobs filled in the City of London (CoL) from which borough does the largest number of workers commute?
- For jobs filled in Camden (Cmd) from which borough does the largest number of workers commute?
- Eyeballing the graphic, identify the top 3 boroughs which appear to have the most localised labour markets in terms of in-commuting.

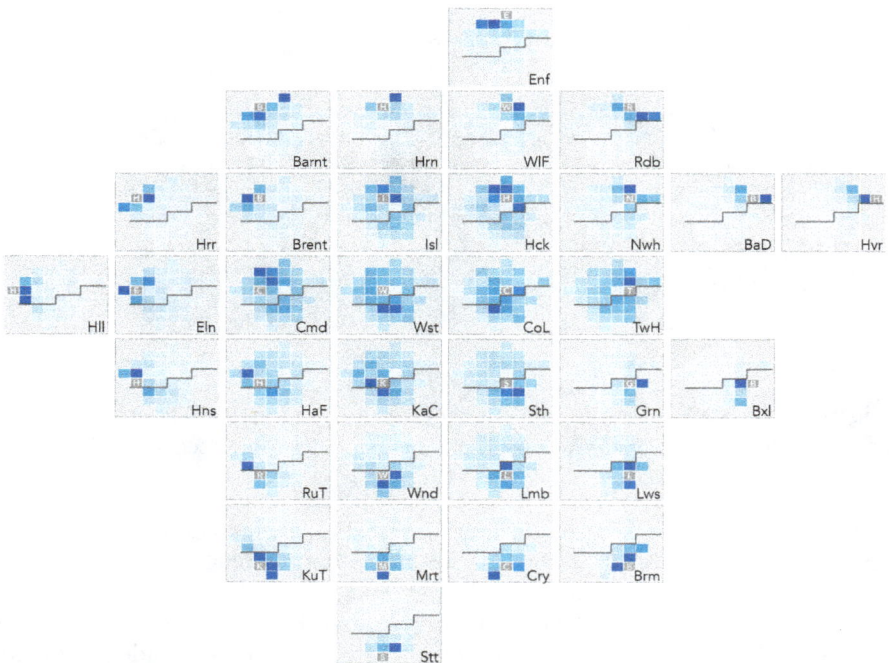

**Figure 5.6:** Destination-focussed OD map of commutes between London boroughs, with local scaling.

## 5.3   Techniques

The technical element to this chapter continues in our analysis of 2011 Census travel-to-work data. After importing the dataset, we will organise flow data into nodes and edges before creating graphics that summarise over the nodes, London boroughs, and reveal spatial structure in the edges, OD commuter flows between boroughs. A focus for the analysis is on how the geography of travel-to-work varies by occupation type.

### 5.3.1   Import

- Download the 05-template.qmd[1] file for this chapter and save it to your vis4sds project.
- Open your vis4sds project in RStudio and load the template file by clicking File > Open File ... > 05-template.qmd.

---

[1] https://vis4sds.github.io/vis4sds/files/05-template.qmd

**Table 5.1:** Census OD travel-to-work data: edges (OD flows) table.

| o_bor | d_bor | occ_type | count | is_prof |
|-------|-------|----------|-------|---------|
| Barnet | Westminster | 1_managers_senior | 2733 | TRUE |
| Barnet | Westminster | 2_professional | 4055 | TRUE |
| Barnet | Westminster | 3_associate_professional | 2977 | TRUE |
| Barnet | Westminster | 4_administrative | 2674 | FALSE |
| Barnet | Westminster | 5_trade | 687 | FALSE |
| Barnet | Westminster | 6_caring_leisure | 755 | FALSE |
| Barnet | Westminster | 7_sales_customer | 1255 | FALSE |
| Barnet | Westminster | 8_machine_operatives | 257 | FALSE |
| Barnet | Westminster | 9_elementary | 1309 | FALSE |
| ... | ... | ... | ... | ... |

A .csv file containing Census travel-to-work data in London has been stored in the book's accompanying data repository[2]. Code for downloading the data is in the template file. The data can then be read into your session in the usual way.

```
# Read in local copies of the Census travel-to-work data.
od_pairs <- read_csv(here("data", "london_ttw.csv"))
```

In order to generate an approximate geographic arrangement of London boroughs we will use the gridmappr R package (Beecham 2024). The development version can be downloaded with:

```
devtools::install_github("rogerbeecham/gridmappr")
```

The od_pairs dataset is in Table 5.1. Each observation is a unique OD pair summarising the total number of recorded commuters between a pair of London boroughs for a stated occupation type.

Nodes in the dataset are the 33 London boroughs. We can express commuters between these nodes in different ways, according to whether nodes are destinations or origins. In the code below, two tables are generated with OD data grouped by destination (nodes_d) and origin (nodes_o) and commuters into- and out of- boroughs counted respectively. These two data sets are then combined with bind_rows() and distinguished via the variable name type.

```
nodes_d <- od_pairs |>
  group_by(d_bor, occ_type) |>
  summarise(count = sum(count), is_prof = first(is_prof)) |>
```

---

[2]https://github.com/vis4sds/data

```
  ungroup() |> rename(la = d_bor) |>
  mutate(type="jobs")

nodes_o <- od_pairs |>
  group_by(o_bor, occ_type) |>
  summarise(count = sum(count), is_prof = first(is_prof)) |>
  ungroup() |> rename(la = o_bor) |>
  mutate(type="workers")

nodes   <- bind_rows(nodes_o, nodes_d)
```

## 5.3.2   Gridmap layout

We will analyse over the travel-to-work data by laying out data graphics with a geospatial arrangement. Such arrangements can be automatically created using the gridmappr R package (Beecham 2024). Given a set of point locations, the package creates a two-dimensional grid of user-specified dimensions and allocates points to the grid such that the distance between points is minimised.

The main function to call is points_to_grid(). This takes a data frame of geographic points and returns corresponding grid cell positions (*row* and *column* identifiers). In the code below an 8x8 grid is used. The allocation is also constrained by a *compactness* parameter which determines the extent to which points are allocated to cells in the centre (compactness = 1), edges (0) or scaled geographic location (0.5) within the grid.

```
library(gridmappr)
n_row <- 8
n_col <- 8
pts <- london_boroughs |> st_drop_geometry() |>
  select(area_name, x = easting, y = northing)
solution <- points_to_grid(pts, n_row, n_col, compactness = .6)
```

Once a layout is generated, we create a corresponding polygon object so that the gridmap can be plotted. This is achieved with make_grid(). This function takes an sf data frame containing polygons with 'real' geography and returns an sf data frame representing a grid, with variables identifying column and row IDs (bottom left is origin) and geographic centroids of grid squares. The gridded object can then be joined on a gridmap solution returned from points_to_grid() in order to create an object in which each grid square corresponds to a gridmap cell position.

```
grid <- make_grid(london_boroughs, n_row, n_col) |>
  inner_join(solution)
```

To evaluate different layouts that could be generated from differently specified grid dimensions and/or compactness values, it can be useful to show the geographic distortion introduced when moving centroids to regularly sized grid cells. In the example below, displacement vectors are drawn connecting the centroid of each borough in London in *real* and *grid* space. This is achieved with `get_trajectory()` from the `odvis` package.

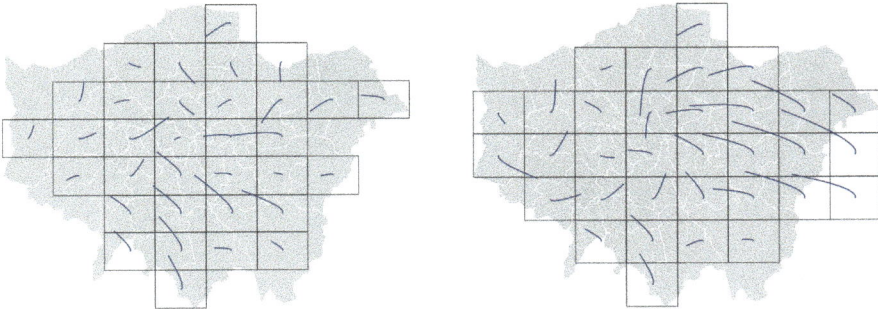

**Figure 5.7:** Displacement vectors showing distortion introduced by candidate gridmap layouts.

The code is slightly more advanced. Some concepts, for example functional-style programming with `map()`, are introduced properly in later chapters. First, we combine the real and grid geographies in a single data frame. Then we `map()` over each real-to-grid location pair calling `get_trajectory()` to generate a data frame of trajectories – origins, destinations and control points, which affect the path of the vectors so that they curve towards the destination. Finally trajectories are plotted via `geom_bezier()`, with separate lines (`group=`) for each real-to-grid OD pair.

```
# Install odvis.
devtools::install_github("rogerbeecham/odvis")
library(odvis)

# Combine the grid and london_boroughs (real geography)
# objects into a single simple features data frame.
lon_geogs <- bind_rows(
  london_boroughs |> mutate(type = "real") |>
    select(area_name, x = easting, y = northing, type),
  grid |>  mutate(type = "grid") |>
```

```r
    select(area_name, x, y, type, geometry = geom)
)

# Create points for drawing trajectories
# -- origin, destination and control point locations.
trajectories <- lon_geogs |> st_drop_geometry() |>
  filter(!is.na(area_name)) |>
  pivot_wider(names_from = type, values_from = c(x, y)) |>
  mutate(id = row_number()) |>
  nest(data = c(area_name, x_real, y_real, x_grid, y_grid)) |>
  mutate(trajectory = map(data,
    ~get_trajectory(
      .x$x_real, .x$y_real, .x$x_grid, .x$y_grid, .x$area_name
      ))
    ) |>
  select(trajectory) |>
  unnest(cols = trajectory)

# Plot displacement vectors.
ggplot() +
  geom_sf(
    data = lon_geogs |>
      mutate(type = factor(type, levels = c("real", "grid"))),
    aes(fill = type, colour = type), linewidth = .2) +
  ggforce::geom_bezier(
    data = trajectories,
    aes(x = x, y = y, group = od_pair),
      colour = "#08306b", linewidth = .4
    ) +
  scale_fill_manual(
    values = c("#f0f0f0", "transparent"), guide = "none") +
  scale_colour_manual(
    values = c("#FFFFFF", "#525252"), guide = "none") +
  theme_void()
```

Once a gridmap polygon file (`grid`) and corresponding cell positions (`row` and `col`) are generated, gridmaps can be plotted from the polygon file directly, as in plot *(a)* of Figure 5.8, by supplying grid square positions to `facet_grid()`, plot *(b)*, or combining both to effect a map-within-map layout for OD maps, plot *(c)*.

(a) Gridmap polygons

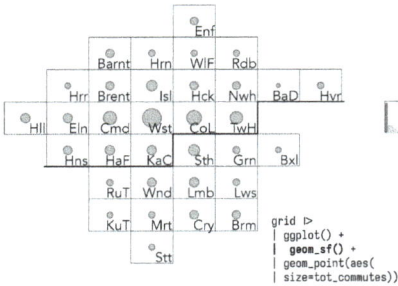

(b) Facet grid

```
grid ▷
| ggplot() +
|  geom_sf() +
|  geom_point(aes(
|   size=tot_commutes))
```

```
edges ▷
| ggplot() +
| geom_col(aes(
|  x=bor_rank, y=commutes)) +
|  facet_grid(-d_row-d_col)
```

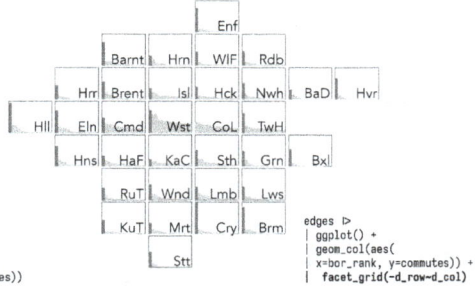

(c) Polygons (origins) + facets (dests)

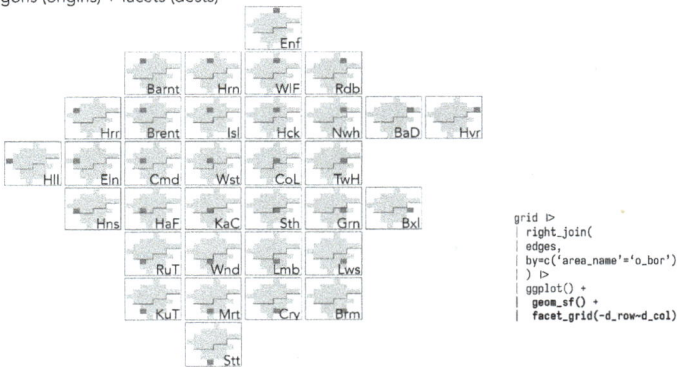

```
grid ▷
| right_join(
| edges,
| by=c('area_name'='o_bor')
| ) ▷
| ggplot() +
|  geom_sf() +
|  facet_grid(-d_row-d_col)
```

**Figure 5.8:** ggplot2 code for plotting gridmap layouts.

### 5.3.3 Analysing over nodes

In Figure 5.9 are gridmaps summarising over the nodes (boroughs). The number of workers living in each borough (left column) and jobs available in each borough (right column) is encoded using circle size, with circles positioned in $x$, $y$ at the centroids of the grid squares. Frequencies are shown separately for *professional* and *non-professional* occupation types. If you are familiar with London's social geography, the patterns can be understood. There are comparatively more non-professional workers living in the somewhat more affordable boroughs in outer and east London; and job-rich central London boroughs – Westminster Wst, Camden Cmd, City of London CoL, Tower Hamlets TwH – provide a large number of *professional* jobs.

The code for Figure 5.9:

```
grid |>
  inner_join(nodes |> group_by(la, is_prof, type) |>
    summarise(count=sum(count)),
```

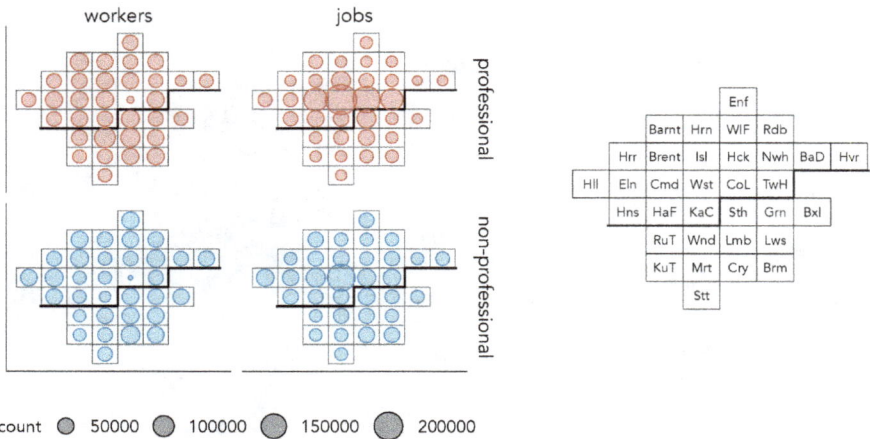

**Figure 5.9:** Workers and jobs in London borough by occupation class. Bar counts are scaled locally by borough.

```
    by = c("area_name" = "la")
  ) |>
mutate(
  is_prof =
  factor(if_else(is_prof, "professional", "non-professional"),
    levels = c("professional", "non-professional")),
  type = factor(type, levels = c("workers", "jobs")),
) |>
ggplot(aes(x = x, y = y)) +
geom_sf(fill = "#ffffff") +
geom_point(aes(size = count, colour = is_prof), alpha = .5) +
facet_grid(is_prof ~ type) +
scale_colour_manual(values = c("#67000d", "#08306b"))
```

The `ggplot2` code:

1. *Data*: From the derived `nodes` data frame we count workers and jobs
   (`type`) by borough, collapsed over professional or non-professional
   occupation types (`is_prof`). Note that we also start by joining on
   `grid` in order to bring in the polygon file and coordinates of the
   generated gridmap. Converting `is_prof` and `type` to factor variables
   gives us control over the order in which they appear in the plot.
2. *Encoding*: the proportional symbols are positioned at the centroids
   of borough grid squares $(x, y)$, sized according to count of jobs or
   workers and coloured according to occupation type (`is_prof`).

3. *Marks*: `geom_point()` for proportional symbols and `geom_sf()` for grid outline – remember our dataset is now of class `sf` as we joined on the `grid` object.
4. *Scale*: `scale_colour_manual()` for associating occupation type.
5. *Facets*: `facet_wrap()` on workers/jobs summary type and high-level occupation type (`is_prof`).

In Figure 5.9, we collapsed over nine occupation types in order to plot proportional-symbol maps. Since gridmaps consist of regularly-sized cells, we can introduce more complex graphical summaries with a geographical arrangement. For example, Figure 5.10 uses bar charts to analyse the number of workers (left-pointing bars) and jobs (right-pointing bars) by occupation type across the the full nine occupation classes. In the selected examples below, jobs and workers are differentiated by varying the direction of bars: pointing to the right for jobs, to the left for workers. The counts are scaled locally. For each borough, its modal category count of jobs/workers by occupation is found, and bar length is scaled relative to this modal category. This encoding allows us to distinguish between job-rich boroughs with longer bars pointing to the right (Westminster); resident/worker-rich boroughs with longer bars pointing to the left (Wandsworth); and outer London boroughs that are more self-contained (Hillingdon).

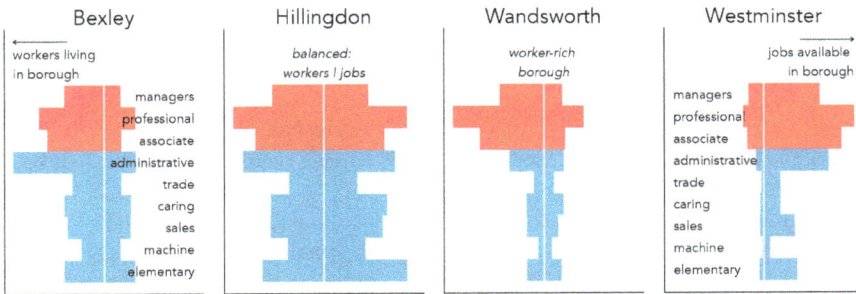

**Figure 5.10:** Workers and jobs in selected London boroughs by full occupation classes.

The code for Figure 5.10:

```
plot_data <- solution |>
  inner_join(nodes, by = c("area_name" = "la")) |>
  group_by(area_name) |>
  mutate(count = count / max(count)) |> ungroup() |>
  mutate(
    count = if_else(type == "jobs", count, -count),
```

```
  occ_name = factor(occ_type),
  occ_type = as.numeric(fct_rev(factor(occ_type)))
)

plot_data |>
  filter(area_name %in%
    c("Wandsworth", "Westminster", "Bexley", "Hillingdon")) |>
  ggplot(aes(x = occ_type, y = count)) +
  geom_col(aes(fill = is_prof), alpha = .5, width = 1) +
  geom_hline(yintercept = 0, linewidth = .4, colour = "#ffffff") +
  facet_wrap(~area_name) +
  scale_y_continuous(limits = c(-1, 1)) +
  scale_fill_manual(values = c("#08306b", "#67000d"), guide = "none") +
  coord_flip()
```

The ggplot2 spec:

1. *Data*: We create a staged dataset for plotting (`plot_data`). The different bar directions for workers/jobs is achieved by a slight hack – changing the polarity of counts by occupation depending on the summary `type`. Additionally in this staged dataset, counts are further locally (borough-level) scaled. We `group_by` borough and express counts of jobs or workers in a borough for an occupation type relative to the largest occupation type in that borough. Note that we `filter()` on some selected boroughs.

2. *Encoding*: Bars whose length (`y=`) varies according to `count` and categorical position (`x=`) according to `occ_type`, filled on high-level occupation type (`is_prof`).

3. *Marks*: `geom_col()` for bars.

4. *Scale*:   `scale_fill_manual()`   for   associating   occupation   type, `scale_x_continuous()` for making sure workers/jobs bars use the same scale.

5. *Facets*: `facet_wrap()` on borough (`area_name`).

6. *Setting*: `coord_flip()` for bars that are oriented horizontally.

Adding a geospatial arrangement, as in Figure 5.11, can further help with exploring the geography to these different categories of borough: balanced boroughs to the east (Barking and Dagenham BaD) and west (Hillingdon Hil); worker-rich boroughs (left-pointing bars) with large proportions of professional workers in west and south west London (Wandsworth Wnd, Richmond Upon Thames RuT); job-rich boroughs (right-pointing bars) in central London (Westminster Wst, Camden Cmd).

Different from the proportional-symbol maps, the spatial arrangement in Figure 5.11 is generated using ggplot2's in-built faceting rather than a spatial

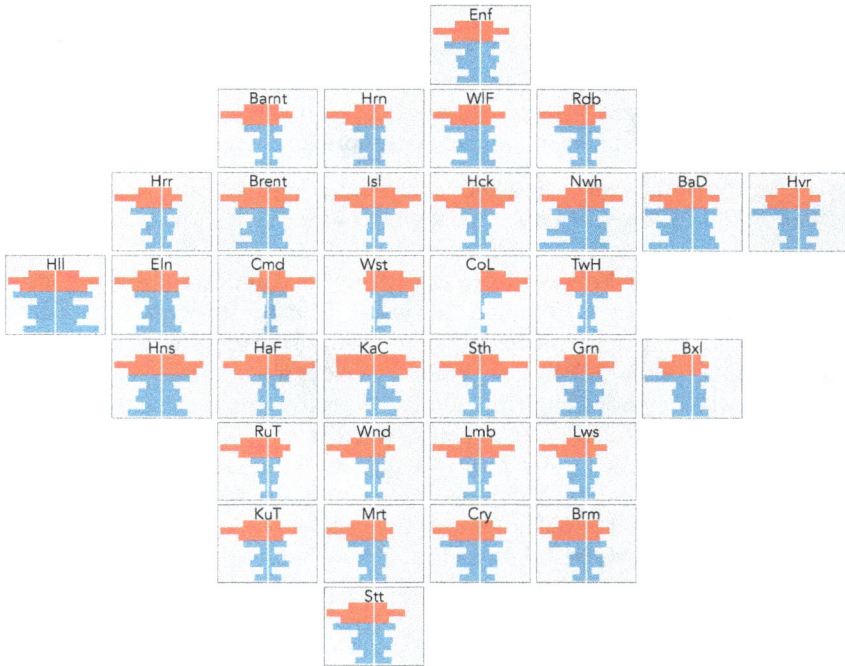

**Figure 5.11:** Workers and jobs in London boroughs by full occupation classes.

polygon file. This can be understood when remembering that gridmap layouts created by `points_to_grid()` define *row* and *column* identifiers for each spatial unit. The only update to the bar chart code is that we supply *row* and *col* identifiers to `facet_grid()`, with a slight hack on the *row* variable (`-row`) as gridmappr's origin *[min-row, min-col]* is the bottom-left cell in the grid whereas for `facet_grid()` the origin is the top-left.

The code for Figure 5.11 is below, simply updating the early code with a call to `facet_grid()`:

```
plot_data |>
  ggplot(aes(x = occ_type, y = count)) +
  geom_col(aes(fill = is_prof), alpha = .5, width = 1) +
  geom_hline(yintercept = 0, linewidth = .4, colour = "#ffffff") +
  facet_grid(-row ~ col, scales = "free") +
  scale_y_continuous(limits = c(-1, 1)) +
  scale_fill_manual(values = c("#08306b", "#67000d")) +
  coord_flip()
```

### 5.3.4    Analysing over edges

To study the geography of flows between boroughs, we can update our ggplot2 specification to generate a full OD map. In the example in Figure 5.12, there is a little more thinking around patterns in the data that we wish to explore, borrowing from the ideas introduced in the previous chapter.

We've identified differences in where professional jobs and workers are located in London, and it is reasonable to expect that flows between boroughs also have an uneven geography. To explore this, we can set up a model that assumes that commuter flows between boroughs distribute uniformly across London. Of all commutes between London boroughs, 51% are to access *professional* jobs (global_prof). Under an assumption of uniformity, were we to randomly sample an OD (borough-borough) commute pair, we would expect to see this proportion when counting up the number of professional and non-professional occupation types present in that commute. For each OD pair, we therefore generate expected counts by multiplying the total number of commuters present in an OD pair by this global_prof, and from here signed residuals (resid) identifying whether there are greater or fewer professionals commuting that OD pair than would be expected. Note that these are like the signed chi-scores  in the previous chapter in that rather than expressing differences in observed counts as a straight proportion of expected counts (dividing by expected counts), we apply a power transform that is <1.0 to the denominator. This has the effect of also giving saliency to differences that are large in absolute terms. You could try varying this exponent (maybe between 0.5-1.0) to see its effect on residuals in the OD map.

Figure 5.12 is a D-OD map; the large reference cells are destination boroughs (workplaces), and the small cells origins (residences) from which workers travel to access jobs in the reference cell. From this we observe that job-rich boroughs in central London are associated more with professional occupations (red cells) and draw professional commuters especially from 'residential' boroughs such as Wandsworth (Wnd), Hammersmith and Fulham (HaF). Note that the darker colours indicate that these job-rich boroughs also attract workers in large number from boroughs across London. By contrast, boroughs in outer London do not draw workers from across London in such large number, and the very dark blues in the reference cells suggest that, as might be expected, the labour market for non-professional jobs is more localised.

The code:

```
edges <- od_pairs |>
  group_by(o_bor, d_bor)  |>
  summarise(
    commutes = sum(count),
    is_prof = sum(count[is_prof]),
```

Expectation: commutes by occupation
*uniformly* distributed

% professional jobs in a
sampled OD pair
~ London total (51%)

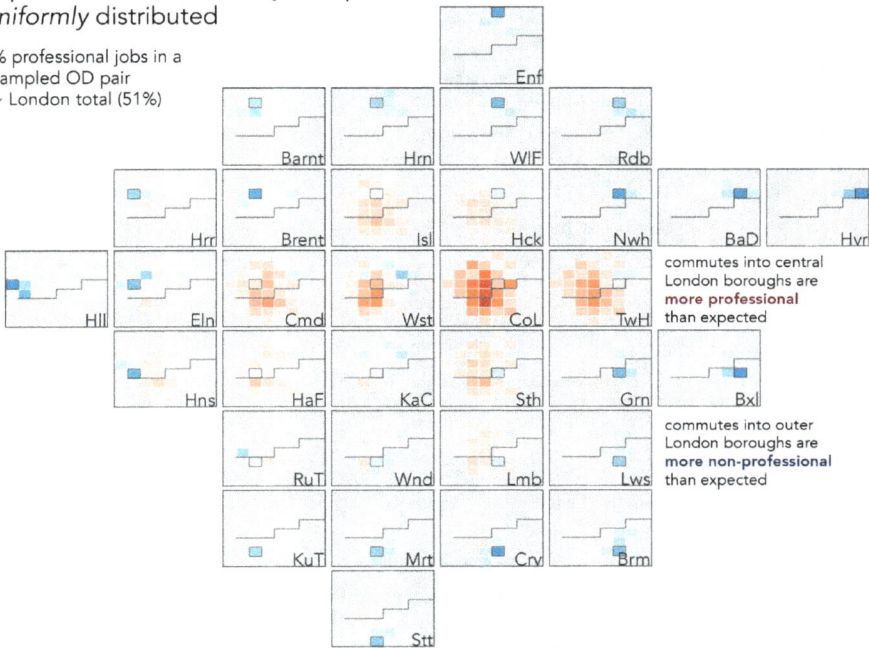

commutes into central
London boroughs are
**more professional**
than expected

commutes into outer
London boroughs are
**more non-professional**
than expected

**Figure 5.12:** Commutes between London boroughs: difference maps by occupation type assuming *professionals* and *non-professionals* distribute uniformly across London.

```
    prop_prof= is_prof/commutes
    ) |>
  left_join(grid, by=c("o_bor"="area_name")) |>
    st_drop_geometry() |> select(-geom) |>
  rename(o_x=x, o_y=y, o_col=col, o_row=row) |>
  left_join(grid, by=c("d_bor"="area_name")) |>
    st_drop_geometry() |> select(-geom) |>
  rename(d_x=x, d_y=y, d_col=col, d_row=row)

plot_data <- edges |>
  mutate(
    non_prof = commutes-is_prof,
    prof = is_prof,
    global_prof = sum(prof) / sum(prof + non_prof),
    count = prof + non_prof,
    obs = prof,
```

```
      exp = (global_prof * count),
      resid = (obs - exp) / (exp^.7)
      ) |>
    # Join on d_bor for an O-OD map.
    left_join(grid |> select(area_name), by = c("o_bor" = "area_name")) |>
    mutate(
      bor_label = if_else(o_bor == d_bor, d_bor, ""),
      bor_focus = o_bor == d_bor
    ) |>
    st_as_sf()

bbox_grid <- st_bbox(grid)
max_resid <- max(abs(plot_data$resid))

plot_data |>
  ggplot() +
  geom_sf(aes(fill=resid), colour = "#616161",
    size = 0.15, alpha = 0.9) +
  geom_sf(data = . %>% filter(bor_focus),
    fill = "transparent", colour = "#373737", size = 0.3
    ) +
  geom_text(
    data = plot_data %>% filter(bor_focus),
    aes(x = bbox_grid$xmax, y = bbox_grid$ymin,
      label = abbreviate(o_bor, 3)),
    colour = "#252525", alpha = 0.9, size = 3.5,
    hjust = "right", vjust = "bottom"
  ) +
  coord_sf(crs = st_crs(plot_data), datum = NA) +
  facet_grid(-d_row ~ d_col, shrink = FALSE) +
  scale_fill_distiller(palette = "RdBu", direction = -1,
    limits=c(-max_resid, max_resid))
```

The `ggplot2` spec:

- Data:
  - Calculate the proportion of professional jobs in the dataset (`global_prof`).
  - Then for each destination (workplace) borough calculate the expected number of commutes for any OD pair by multiplying the number of jobs contained in that OD pair by `global_prof`, and express the difference between the actual number of professional jobs as a rate with a power transform (`(obs-exp) / (exp^.7)`).
  - Take the staged dataset, and join twice on the `gridmap` dataset.
  - Then join the with the gridded polygon file (`grid`) on `o_bor` – in this OD map the small cells are origins.

- Finally, in the mutate() we generate a new variable identifying the borough in focus (bor_focus), destination in this case, and a text label variable for annotating plots (bor_label).
- Encoding:
  - Gridmap cells are coloured according to the calculated residuals (fill=resid).
  - Text labels for focus (workplace) boroughs are drawn in the bottom-right corner of larger cells. Note that the coordinate space here is that from the gridmap dataset, and so the $x,y$ location of borough labels is derived from the bounding box object (bbox_grid), calculated during data staging.
- Marks: geom_sf() for drawing the small grid-cell maps; geom_text() for drawing the labels.
- Scale: scale_fill_distiller() for a diverging colour scheme using the Color-Brewer (Harrower and Brewer 2003) RdBu palette and made symmetrical on 0 by manually setting limits() based on the maximum residual value.
- Facets: facet_grid() for effecting the map-within-map layout.

Once the data staging and ggplot2 code for the OD map is generated, it is very easy to adapt and extend the code to explore different assumptions. For example, the expectation of a uniform distribution across London in the relative number of commutes by occupation type is a flawed one since we know that there is some variation in the proportion of professional jobs available in each borough. In the City of London (CoL) 74% of jobs are professional whereas in Bexley (Bxl), Havering (Hvr) and Barking and Dagenham (BaD), that figure is c.30%. We can easily adapt the data staging code to instead generate local expectations for each destination borough by moving the assignment of global_prof into a group_by() on destination borough. The expectation is now that the relative number of professional commutes present in any OD pair should be proportionally equivalent to the number of professional jobs available at that OD pair's destination borough. Colouring cells of the OD map according to this new quantity (Figure 5.13) exposes patterns that relate to London's social geography: greater than expected *non-professional* workers from more affordable boroughs to the east of London and into job-rich boroughs in central London and a reverse pattern for origin boroughs supplying greater than expected *professional* workers.

## 5.4 Conclusions

Network data are challenging to represent, work with and analyse. It is for this reason that visual approaches are often used in their analysis. A common pitfall to many network visualizations is that they simply re-present that complexity

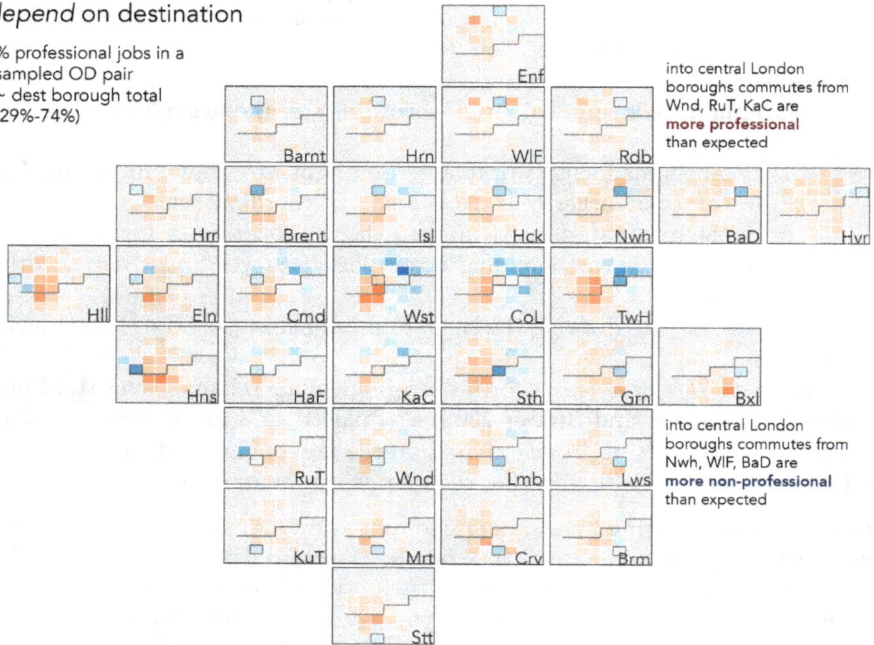

**Expectation: commutes by occupation *depend* on destination**

% professional jobs in a sampled OD pair ~ dest borough total (29%-74%)

into central London boroughs commutes from Wnd, RuT, KaC are **more professional** than expected

into central London boroughs commutes from Nwh, WlF, BaD are **more non-professional** than expected

**Figure 5.13:** Commutes between London boroughs: difference maps by occupation type assuming *professionals* and *non-professionals* distribute uniformly *within* boroughs.

without exposing useful structure or insight into the phenomena being analysed. Through an analysis of 2011 Census travel-to-work data in London, this chapter demonstrated approaches to analysing and inferring structure in a category of network data common to geographers: geospatial origin-destination data. Spatially-arranged node-link diagrams are highly intuitive and can support a kind of synoptic overview of a network, but were of limited success in representing detailed patterns in travel-to-work within and between London boroughs. Instead we used matrix-based views, including spatially arranged matrices or OD Maps. As ever, the appropriateness of either approach, node-link based or matrix-based representations, depends on data, analysis purpose and audience.

## 5.5 Further Reading

For working with network data in tidyverse and ggplot2:

- Wickham, H., Navarro, D. and Lin Pedersen, T. 2023. "ggplot2: Elegant Graphics for Data Analysis Third Edition.", New York, NY: *Springer*.

For the original OD maps paper:

- Wood, J., Dykes, J. and Slingsby, A. 2010. "Visualisation of Origins, Destinations and Flows with OD Maps." *The Cartographic Journal*, 47(2): 117–29. doi: 10.1179/000870410X12658023467367.

Not about network visualization per se, but presents numerous (100!) data graphics on London. Worth highlighting here is the use of annotations and efficient graphical descriptions, a theme we return to in Chapter 8:

- Cheshire, J. and Uberti, O. 2016 "London, The Information Capital: 100 maps and graphics that will change how you view the city", London, UK: *Penguin*.

# 6

Models

By the end of this chapter you should gain the following knowledge and practical skills.

---

**i Knowledge**

☐ Be reminded of the basics of linear regression modelling.
☐ Appreciate how data graphics can inform the process of building and evaluating models.
☐ Understand two categories of geographic effect in regression modelling: spatial *dependence* in values and spatial *non-stationarity* in processes.
☐ Learn how graphics can be used to test for these effects and how linear regression models can be updated to account for and further explore them.

---

**i Practical skills**

☐ Write ggplot2 code to generate graphics for exploring multivariate association and presenting regression outputs (faceted scatterplots, parallel coordinate plots, dot plots with error bars).
☐ Write code to generate linear regression models in R.
☐ Extract model outputs and diagnostics in a `tidy` manner.
☐ Apply functional-style programming for working over multiple model outputs.
☐ Generate graphical line-up plots in ggplot2 to test regression assumptions.

---

## 6.1 Introduction

So far the analysis presented in this book has been data-driven. Having described data in a consistent way (Chapter 2), visual analysis approaches have been applied, informed by established visualization guidelines. Chapters 4 and 5 involved model building, but these were largely value-free models based on limited prior theory. This chapter is presented as a worked data analysis. We look at a well-known dataset with a more explicit and theoretically-informed motivation.

The chapter explores variation in voting behaviour in the UK's 2016 referendum on leaving the EU. You might remember that while there was a slight majority for Leave (c. 52%), the vote varied between different parts of the country. There were many explanations offered for why particular places voted the way they did, often related to the demographic composition of those areas. We will explore whether the discussed compositional demographic factors vary systematically with area-level Leave voting. Using regression frameworks, we will model the relative effect of each of these compositional factors in structuring variation in the vote and construct data graphics that allow these models and parameters to be evaluated in detail.

> **i** Regression primer
>
> This chapter assumes some basic familiarity with linear regression modelling. For a fuller overview, with excellent and real-world social science examples, you may wish to consult *Regression and Other Stories* (Gelman, Hill, and Vehtari 2020).

## 6.2   Concepts

### 6.2.1   Quantifying and exploring variation

In Figure 6.1 is a map and bar chart of voting in the 2016 EU referendum, estimated at Parliamentary Constituency level (see Hanretty 2017). The values themselves are the difference in estimated vote shares from an expectation that the Leave vote by constituency, $y_i$ our *outcome* of interest, is uniformly distributed across the country and so equivalent to the overall Great Britain (GB) vote share for Leave of c. 52%. Although a slightly contrived formulation, we could express this as an intercept-only linear regression model, where the estimated *slope* ($\beta_1$) is 'turned off' (takes the value 0) and the *intercept* ($\beta_0$) is the GB average vote share for Leave ($\bar{y}$):

$$y_i = \beta_0 + \beta_1 + \varepsilon_i$$

So we estimate the Leave vote in each constituency ($y_i$) as a function of:

- $\beta_0$, the intercept, the GB average vote share ($\bar{y}$) +
- $\beta_1 = 0$, a negated slope, +

- $\varepsilon_i$, a statistical error term capturing the difference between $y_i$, the observed Leave vote in a constituency, and the unobservable 'true population' value of the Leave vote in each constituency

How does this relate to the idea of characterising variation? The length and colour of each bar in Figure 6.1 is scaled according to model *residuals*: the difference between $y_i$, the observed value, and the expected value of the Leave vote under the uniform model. The sum of these bar lengths is therefore the total variance that we later try to account for by updating our regression model to generate new expected values using information on the demographic composition of constituencies.

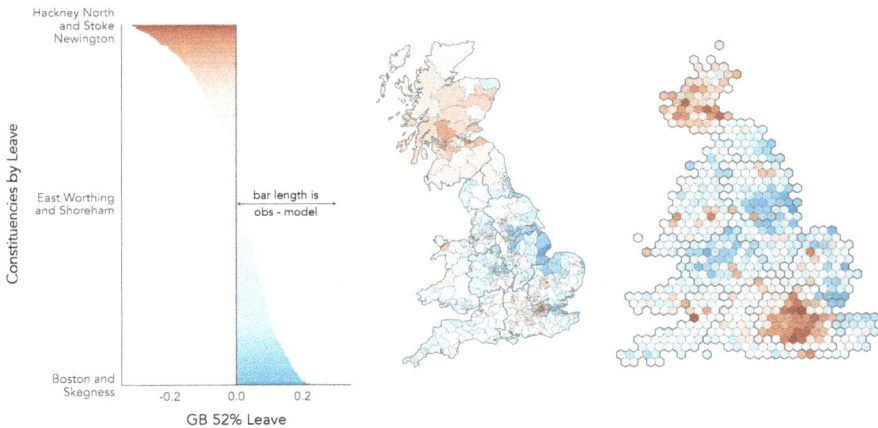

**Figure 6.1:** Residuals from uniform model comparing constituency Leave vote to GB average.

Figure 6.1 is similar to the maps that were published widely in press reports in the aftermath of the vote, and demonstrates that there is indeed substantial variation in Leave voting between different parts of the country. The intercept-only model consistently underestimates the vote in Scotland and most of London. Outside of this, constituencies voting in smaller proportions than would be expected for Leave are distributed more sparsely in the country: the dark red dot with surrounding red area in the east of England is Cambridge and Cambridgeshire, constituencies in Bristol (south west), Manchester and Liverpool (north west) and Brighton (south) are also reasonably strong red.

When evaluating the effectiveness of modelled values, there are various checks that can be performed. A relevant check here is whether there is bias in the residuals – whether residuals have structure that suggests they are grouped in a way not captured by the model. Given the motivation behind our analysis, it is no surprise that there is a geographic pattern to the residuals in Figure 6.1, but also the non-symmetrical shape of the 'signed' bars in the left of the

**Table 6.1:** Breakdown of variable types.

| Census variable | Constituency % |
|---|---|
| post-industrial / knowledge economy | |
| degree-educated | with degrees + |
| professional occupations | ns-sec manager/professional |
| younger adults | adults aged <44 |
| heavy industry | manufacturing and transport |
| diversity/values/outcomes | |
| not good health | reported fair, bad, very bad |
| white | ethnicity white British/Irish |
| Christian | Christian |
| EU-born | EU-born (not UK) |
| metropolitan / 'big city' | |
| own home | own home |
| no car | don't own a car |

graphic. There are more constituencies with positive values than negative; the Leave vote is underestimated by the uniform model for 57% of constituencies, and some constituencies have quite large negative values. The strongest vote for Leave was Boston and Skegness with 76% for Leave, but the strongest for Remain was Hackney North and Stoke Newington with 80% for Remain.

## 6.2.2 Quantifying and exploring co-variation

More interesting still is whether the pattern of variation in Figure 6.1 is correlated with compositional factors that we think explain this variation; and whether bias or structure in residuals exists even after accounting for these compositional factors.

In Table 6.1 is a list of possible explanatory variables describing the demographic composition of constituencies. Each variable is expressed as a proportion of the constituency's population. So the *degree-educated* variable describes the proportion of residents in the constituency educated at least to degree-level. Comparison across these variables is challenging due to the fact that their ranges differ: the *EU-born* variable ranges from 0.6% to 17%; the *white* variable from 14% to 98%. Common practice for addressing these sorts of range problem is to z-score transform the variables so that each is expressed in standard deviation units from its mean.

Figure 6.2 presents scatterplots from which the extent of linear association between these demographics and Leave voting in each constituency can be inferred. Each dot is a constituency, arranged on the x-axis according to the value of the explanatory variable and the y-axis according to the share of Leave vote. The scatterplots are faceted by explanatory variable and ordered left-to-right and top-to-bottom according to correlation coefficient. The variable

most heavily correlated with Leave voting is *degree-education*: as the share
of a constituency's population educated at least to *degree-level* increases, the
share of Leave vote in that constituency decreases. An association in the same
direction, but to a lesser extent, is observed for variables representing similar
concepts: *professional occupations, younger adults, EU-born, no-car* and the
reverse for *Christian, not-good health* and *heavy industry*.

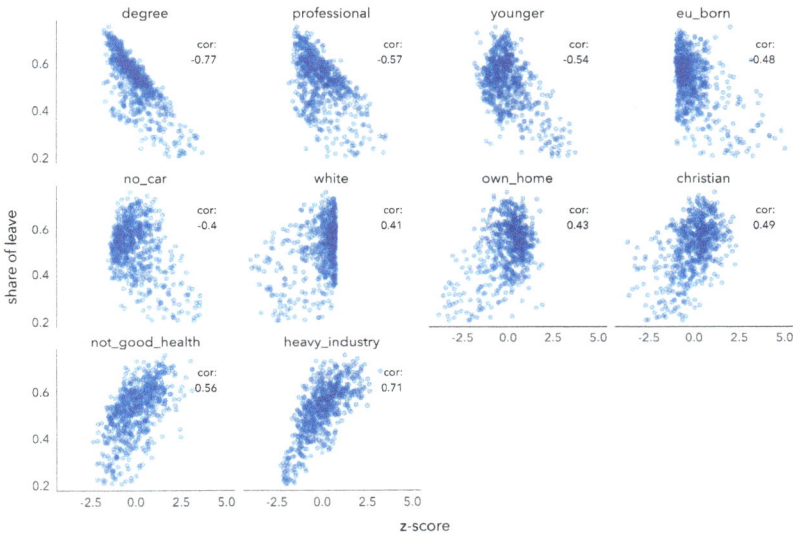

**Figure 6.2:** Scatterplots of constituency Leave vote against selected explanatory variables.

It is of course likely that many of the compositional characteristics of constituencies vary with Leave voting in consistent ways. Parallel coordinate plots
may help visually explore this multivariate space. Whereas in scatterplots
observations are represented as points located in x- and y- axes that are or-
thogonal, in a parallel coordinate plot observations are laid out across many
parallel axes and the values of each observation encoded via a line connecting
the multiple parallel axes. In Figure 6.3 each of the thin lines is a constituency
coloured according to the recorded voting outcome – either majority Remain
(red) or Leave (blue). The first variable encoded is the size of the Leave vote,
and variables are then ordered on their linear association with Leave. Note
that we have reversed the polarity of variables such as *degree-educated* and
*professional* so that we expect more Leave (blue lines) towards the right locations of the parallel axes. That the blue and red lines are reasonably separated
suggests that there is a consistent pattern of association across many of the
demographics characteristics in constituencies voting differently on Leave and
Remain.

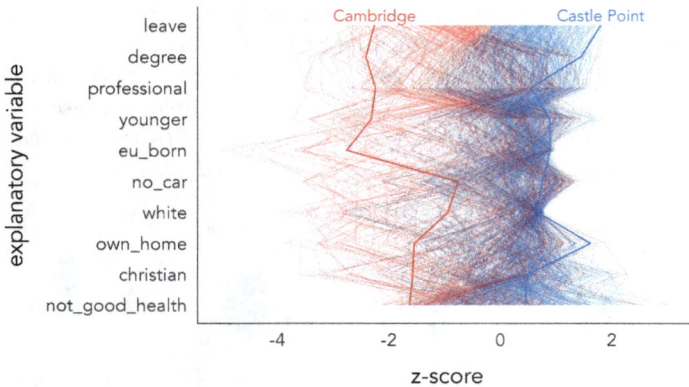

**Figure 6.3:** Parallel coordinate plot of constituency Leave vote and selected explanatory variables.

---

**i On parallel coordinate plots**

Although parallel coordinate plots enable some aspects of association between multiple variables to be inferred, they have several deficiencies. Association can only be directly inferred by comparing variables that are immediately adjacent. The order of parallel variables can greatly affect their visual appearance. And the corollary is that visual patterns of the plot that are salient may be incidental to the statistical features being inferred.

---

### 6.2.3   Modelling for co-variation

Linear regression provides a framework for systematically describing the associations implied by the scatterplots and parallel coordinate plot, and with respect to the constituency-level variation identified in Figure 6.1. Having seen these data, the demographic variables in Figure 6.2, we can derive new expected values of constituency-level Leave voting.

To express this in equation form, we update the uniform model such that Leave vote is a *function* of the selected explanatory variables. For single-variable linear regression, we might select the proportion of residents educated at least to *degree-level* ($d_{i1}$):

$$y_i = \beta_0 + \beta_1 d_{i1} + \varepsilon_i$$

So we now estimate the Leave vote in each constituency ($y_i$) as a function of:

- $\beta_0$, the intercept, the GB average vote share $(\bar{y})$ +
- $\beta_1 = \beta_1 d_{i1}$, the slope, indicating in which direction and to what extent *degree-educated* is associated with Leave, +
- $\varepsilon_i$, the difference between $y_i$ (the observed value) and the unobservable 'true population' value of the Leave vote in that constituency (statistical error)

It is of course likely that some demographic variables account for different elements of variation in the Leave vote than others. You will be aware that the linear regression model can be extended to include many explanatory variables:

$$y_i = \beta_0 + \beta_1 x_{i1} + ... + \beta_k x_{ik} + \varepsilon_i$$

So this results in *separate* $\beta_k$ coefficients for separate explanatory variables. These coefficients can be interpreted as the degree of association between the explanatory variable $k$ and the outcome variable, keeping all the other explanatory variables constant – or the distinct correlation between an explanatory variable $k$ and the outcome variable, net of the other variables included in the model.

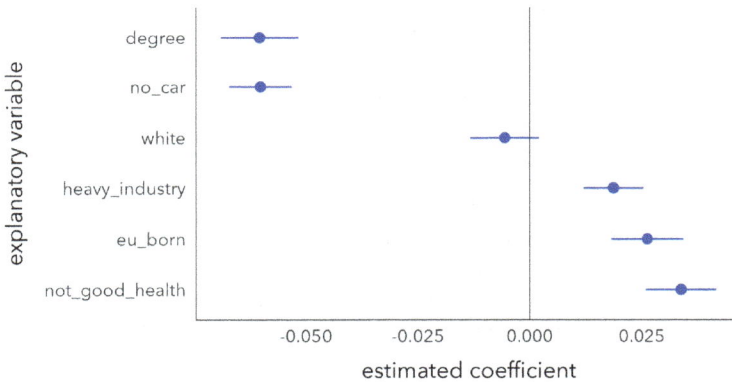

**Figure 6.4:** Outputs from multiple regression model of Leave vote by demographic composition of constituency.

In Figure 6.4 are regression coefficients $(\beta_k)$ from a multiple regression model with *degree-educated, no car, white, heavy industry, EU-born* and *not good health* selected as explanatory variables. Coefficients are reported as dots with estimates of uncertainty represented as lines encoding 95% confidence intervals. Most variables' coefficients are in the direction that would be expected given the associations in Figure 6.2. Net of variation in the other compositional factors, increased levels of *degree-education* in a constituency has the effect of reducing the Leave vote. The two exceptions are *EU-born* and *white*: after controlling for variation in the other demographic variables, increased proportions of *white* residents reduces the Leave vote, and increased proportions of residents that are *EU-born* increases the Leave vote. Since the confidence interval for *white* crosses zero, this coefficient is subject to much uncertainty. Further exploration may allow us to identify whether these counter-intuitive effects are genuine or the result of a poorly-specified model.

### 6.2.4   Evaluating model bias

Our analysis becomes more interesting when we start to explore and characterise model *bias*: any underlying structure to the observations that is less well accounted for by the model.

For area-level regression models such as ours, it is usual for residuals to exhibit some spatial autocorrelation structure. For certain parts of a country a model will overestimate an outcome given the relationship implied between explanatory and outcome variables; for other parts the outcome will be underestimated. This might occur due to:

* *Spatial dependence in variable values* over space. We know that the geography of GB is quite socially distinctive, so it is reasonable to expect, for example, the range in variables like *heavy industry* and *white* to be bounded to economic regions and metropolitan-versus-peripheral regional contexts.
* *Spatial nonstationarity in processes* over space. It is possible that associations between variables might be grouped over space – that the associations vary for different parts of the country. For example, high levels of *EU-born* migration might affect political attitudes, and thus area-level voting, differently in different parts of the country.

We can test for and characterise spatial autocorrelation in residuals by performing a graphical inference test, a map line-up (Beecham et al. 2017; Wickham et al. 2010) against a null hypothesis of *complete spatial randomness* (CSR). A plot of real data, the true map of residuals, is hidden amongst a set of decoys; in this case maps with the residual values randomly permuted around constituencies. If the real map can be correctly identified from the decoys, then this lends statistical credibility to the claim that the observed data are not consistent with the null of CSR. Graphical line-up tests have been used in various domains, also to test regression assumptions (Loy, Hofmann, and Cook 2017). The map line-up in Figure 6.5 demonstrates that there *is* very obviously spatial and *regional* autocorrelation in residuals, and therefore structure that our regression model misses.

There are different ways of updating our model according to this geographic context. We have talked about patterning in residuals as being *spatial*, with values varying smoothly and continuously depending on location. This might be the case, but given the phenomena we are studying, it also plausible that distinct contexts are linked to regions. The residuals in Figure 6.5 – the real being plot 3 – do seem to be grouped by regional boundaries, particularly

1. Analyst observes apparent spatial autocorrelation structure in model residuals.

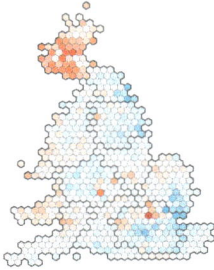

3. If the real is correctly identified, we reject the null of CSR in residuals.

2. Independent observer asked to pick real data from a group of decoys.

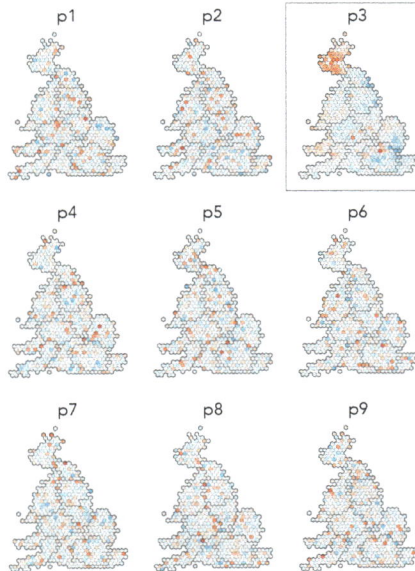

**Figure 6.5:** Map line-up of residuals in which the 'real' dataset is presented alongside 8 decoy plots generated under assumption of CSR.

Scotland looks categorically different. This suggests that geographic context might be usefully represented as a *category* rather than a continuous variable (location in $x,y$). We will therefore update our model representing geographic context as a *regional* grouping and cover approaches both to modelling *spatial dependence* in *values* and *spatial nonstationarity* in *processes*.

## 6.2.5 Geographic context as grouped nuisance term

A common approach to treating geographic dependence in the values of variables is to model geographic context as a Fixed Effect (FE). A dummy variable is created for each group (region in our case), and every region receives a constant. Any group-level sources of variation in the outcome are collapsed into the FE variable, which means that regression coefficients are not complicated by this more messy variation – they now capture the association between demographics and Leave after adjusting for systematic differences in the Leave vote due to region. So, for example, we know that Scotland is politically different from the rest of GB and that this appears to drag down the observed Leave vote for its constituencies. The constant term on region adjusts for this and prevents the estimated regression coefficients (inferred associations between variables) from being affected. Also estimated via the constant is the 'base level' in the outcome for each element of the group – net of demographic composition, the expected Leave vote in each region.

The linear regression model, extended with the FE term ($\gamma_j$), for a single variable model:

$$y_i = \gamma_j + \beta_1 x_{i1} + \varepsilon_i$$

So we now estimate the Leave vote in each constituency ($y_i$) as a function of:

- $\gamma_j$, a constant term similar to an intercept for region $j$, +
- $\beta_1 = \beta_1 x_{i1}$, the slope, indicating in which direction and to what extent some explanatory variable measured at constituency $i$ is associated with Leave, +
- $\varepsilon_i$, the difference between $y_i$ (the observed value) at constituency $i$ and the *unobservable* true population value of the Leave vote in that constituency (statistical error)

Presented in Figure 6.6 are updated regression coefficients for a multivariate model fit with a FE on region. In the left panel are the FE constants. Together these capture the variance in Leave vote between regions after accounting for demographic composition. London is of particular interest. When initially analysing variation in the vote, constituencies in Scotland and London were distinctive in voting in much smaller proportions than the rest of the country for Leave. Given the associations we observe with Leave voting and demographic composition, however, if we were to randomly sample two constituencies that contain the same demographic characteristics, one in London and one in another region (say North West), on average we would expect the Leave vote for the London constituency to be higher (~60%) than that sampled from North West (~51%). A separate and more anticipated pattern is that Scotland would have a lower Leave vote (~38%) – that is, net of demographics there is some additional context in Scotland that means Leave is lower than in other regions.

In the right panel are the regression coefficients net of this between-region variation. Previously the *white* variable had a slight negative association with Leave, counterintuitively. Now the *white* variable has a direction of effect that conforms to expectation – net of variation in other demographics, increased proportions of *white* residents is associated with increased Leave voting. For another variable, *EU born*, the coefficient still unexpectedly suggests a positive association with Leave.

## 6.2.6   Geographic context as grouped effects

Rather than simply allowing a constant term to vary, we can update the linear regression model with an interaction term ($\beta_{1j} x_{i1}$) that permits the coefficient estimates to vary depending on region. This means we get a separate constant term and coefficient estimate of the effect of each variable on Leave for every region.

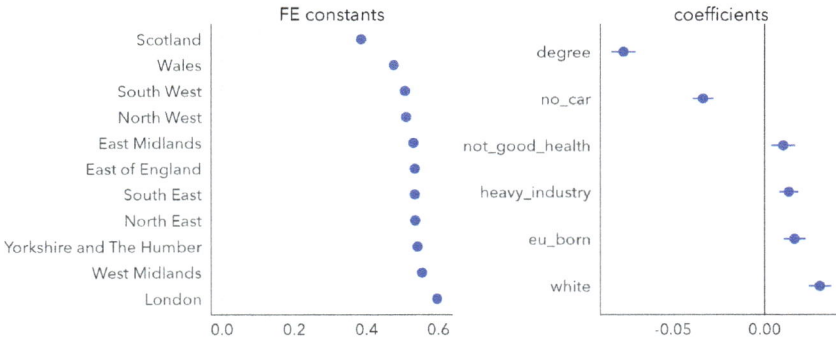

**Figure 6.6:** Output from a multiple regression model of Leave voting against the demographic composition of constituencies, with a Fixed Effect term on region.

$$y_i = \gamma_j + \beta_{1j} x_{i1} + \varepsilon_i$$

- $\gamma_j$, a constant term similar to an intercept for region $j$, +
- $\beta_{1j} x_{i1}$, the region-specific slope, indicating in which direction and to what extent some demographic variable at constituency $i$ and in region $j$ is associated with Leave, +
- $\varepsilon_i$, the difference between $y_i$ (the observed value) at constituency $i$ and the *unobservable* true 'population' value of the Leave vote in that constituency (statistical error)

In Figure 6.7 are region-specific coefficients derived from a multivariate model fit with this interaction term. In each region, *degree-educated* has a negative coefficient and with reasonably tight uncertainty estimates, or at least CIs that do not cross 0. The other variables are subject to more uncertainty. The *no-car* variable is also negatively associated with Leave, a variable we thought may separate metropolitan versus peripheral contexts, but the strength of negative association, after controlling for variation in other demographic factors, does vary by region. The *heavy industry* variable, previously identified as being strongly associated with Leave, has a clear positive association only for London and to a much lesser extent for North West and Wales (small coefficients). The *EU-born* variable is again the least consistent as it flips between positive and negative association when analysed at the regional-level: after controlling for variation in other demographic characteristics, it is positively associated with Leave for North West, Scotland, South West, but negatively associated with Leave for the North East, though with coefficients that are subject to much variation.

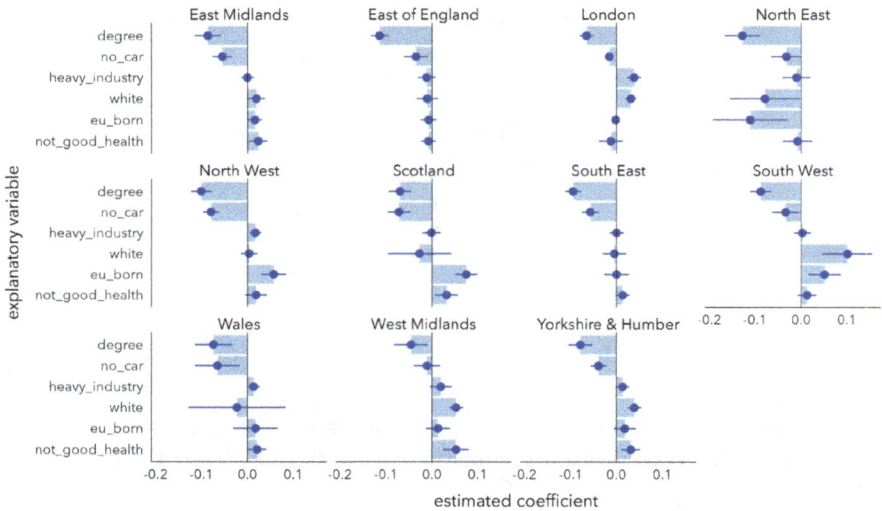

**Figure 6.7:** Output from multiple regression model of Leave vote by demographic composition of constituency with a Fixed Effect and interaction term on region.

### 6.2.7    Estimate volatility and alternative modelling approaches

Our treatment of regression frameworks has in this chapter been reasonably breezy; there are problems that we have not discussed. Introducing FE and interaction terms without adding data reduces statistical power as data are heavily partitioned. Given the fact that our data are hierarchically structured (constituencies sit within regions), hierarchical or multi-level modelling may be more appropriate to this sort of regional grouping. Multi-level modelling uses partial pooling, borrowing data to make estimated coefficients more conservative, less locally biased, where there are comparatively few observations in particular groupings (see Gelman and Hill 2006). There are also many ways in which associations between values can be modelled *continuously* over space. For the case of geographically weighted regression (GWR) (Brunsdon, Fortheringham, and Charlton 2002), local regression coefficients for each spatial unit. Geographically Weighted-statistics enable spatial non-stationarity in process to be flexibly explored and characterised – in this case study, interesting and explainable directions of effect between Leave voting and *EU-born* (see Beecham, Slingsby, and Brunsdon 2018). Since GWR involves generating many hundreds of parameter estimates, visual approaches are as ever primarily used in their interpretation and analysis (see Dykes and Brunsdon 2007).

## 6.3   Techniques

The technical element to this chapter demonstrates how linear regression models can be specified in R, including approaches to extract model summaries and diagnostics, and of course how to represent and evaluate them using data graphics. Data recording estimated vote shares for Leave by Parliamentary Constituency, as well as constituency-level Census demographics, were originally collected from the `parlitools` package.

### 6.3.1   Import, transform, explore

* Download the `06-template.qmd`[1] file for this chapter, and save it to your `vis4sds` project.
* Open your `vis4sds` project in RStudio, and load the template file by clicking `File > Open File ... > 06-template.qmd`.

The template file lists the required packages: `tidyverse`, `sf` and `tidymodels` for extracting model outputs. The processed data with selected 2011 Census demographics can be loaded from the book's accompanying data repository. In this folder is also a `.geojson` file containing a hexagon cartogram of UK parliamentary constituencies, derived from Open-Innovations' `HexJSON` format.

Explanatory variables describing the demographic composition of constituencies are recorded as proportions. In order to support comparison in the multivariate models, they must be z-score transformed. The distance between observed values for each 2011 Census variable is expressed in standard deviation units from the mean across constituencies for that variable. Our approach is to perform this transformation on each explanatory variable before piping into the model specification. This is achieved with `across()`. The first argument is the set of columns to which you would like the same function to be applied, and the second is the function you would like to apply. Remembering that `mutate()` works over columns of a data frame, and that a single column of a data frame is a vector of values, the notation `.x` is used to access each element of the columns being worked across.

```
# z-score transform explanatory variables before model
# specification.
cons_data |>
  mutate(
    across(
      .cols=c(younger:heavy_industry),
      .fns=~(.x-mean(.x))/sd(.x)
```

---

[1]`https://vis4sds.github.io/vis4sds/files/06-template.qmd`

```
      )
   )
   <some-model-specification-code>
```

In Figure 6.2 and Figure 6.3 associations between candidate explanatory variables and Leave are explored using scatterplots and parallel coordinate plots respectively. To avoid cluttering this section, documented code for reproducing these plots is in the `06-template.qmd` file for this chapter and inserted below, but without detailed explanation.

```r
# Data staging and ggplot2 code for PCPs --------------------------

# Pull out and order variable names on their correlation with Leave.
order_vars <- cons_data |>
  mutate(across(c(younger:heavy_industry), ~(.x-mean(.x))/sd(.x))) |>
  pivot_longer(
    cols=younger:heavy_industry, names_to="expl_var", values_to="prop"
    ) |>
  group_by(expl_var) |>
  summarise(cor=cor(leave,prop)) |> ungroup() |> arrange(cor) |>
  pull(expl_var)
# Create staged dataset for plotting.
plot_data <- cons_data |>
  mutate(
    majority=if_else(leave>.5, "Leave", "Remain"),
    across(c(leave, younger:heavy_industry), ~(.x-mean(.x))/sd(.x)),
    decile=ntile(leave, 10),
    is_extreme = decile > 9 | decile < 2
  ) |>
  # Select out variables needed for plot.
  select(
    majority, is_extreme, constituency_name, leave,
    degree, professional, younger, eu_born, no_car, white, own_home,
    christian, not_good_health, heavy_industry
    ) |>
  # Change polarity in selected variables.
  mutate(
    degree=-degree, professional=-professional, younger=-younger,
    eu_born=-eu_born, no_car=-no_car
  ) |>
  # Gather explanatory variables for along rows.
  pivot_longer(
    cols= c(leave:not_good_health), names_to="var", values_to="z_score"
```

```
   ) |>
  # Recode new explanatory variable as factor ordered according to
  # known assocs. Reverse order here as coord_flip() used in plot.
  mutate(
    var=factor(var, levels=c("leave", order_vars)),
    var=fct_rev(var)
  )
# Plot PCP.
plot_data |>
  ggplot(
    aes(x=var, y=z_score, group=c(constituency_name),
    colour=majority)) +
  geom_path( alpha=0.15, linewidth=.2) +
  scale_colour_manual(values=c("#2166ac", "#b2182b")) +
  coord_flip()
```

---

### Task

While the template provides code for reproducing the faceted scatterplots
and parallel coordinate plots, there are some omissions. You will notice
that in Figure 6.3 two very high Leave and Remain constituencies are
highlighted, using thicker red and blue lines and labelling.

Can you update the ggplot2 spec to create a similar effect? There
are different ways of doing this, but you may want to add a separate
`geom_segment()` layer `filter()`-ed on these selected boroughs. The text
annotations may be generated manually using `annotate()`, or derived from
data using `geom_text()`.

---

## 6.3.2   Model tidily

The most straightforward way of specifying a linear regression model is with
the `lm()` function and `summary()` to extract regression coefficients.

```
model <- cons_data |>
  mutate(
    across(
      .cols=c(younger:heavy_industry),
      .fns=~(.x-mean(.x))/sd(.x)
    )
  ) %>%
  lm(leave ~ degree, data=.)
```

```
summary(model)
# Call:
# lm(formula = leave ~ degree, data = .)
#
# Residuals:
#      Min       1Q    Median       3Q       Max
# -0.25521 -0.02548   0.01957   0.05143   0.11237
#
# Coefficients:
#              Estimate Std. Error t value Pr(>|t|)
# (Intercept)  0.520583   0.002896  179.78   <2e-16 ***
# degree      -0.088276   0.002898  -30.46   <2e-16 ***
# ---
# Signif. codes:  0 '***' 0.001 '**' 0.01 '*' 0.05 '.' 0.1 ' ' 1
#
# Residual standard error: 0.07279 on 630 degrees of freedom
# Multiple R-squared:  0.5956,  Adjusted R-squared:  0.595
# F-statistic: 927.9 on 1 and 630 DF,  p-value: < 2.2e-16
```

With `tidymodels`, specifically the `broom` package, we can extract model outputs in a format that adheres to tidy data (Wickham 2014).

- `tidy()` returns estimated coefficients as a data frame.

```
tidy(model)
# # A tibble: 2 × 5
# term         estimate std.error statistic   p.value
# <chr>           <dbl>     <dbl>     <dbl>     <dbl>
# 1 (Intercept)   0.521   0.00290     180.   0
# 2 degree       -0.0883   0.00290     -30.5 5.67e-126
```

- `glance()` returns a single row containing summaries of model fit.

```
glance(model)
# # A tibble: 1 × 12
# r.sq adj.r.sq  sigma     stat p.value    df logLik    AIC    BIC
# <dbl>    <dbl>  <dbl>    <dbl>   <dbl>  <dbl>  <dbl>  <dbl>  <dbl>
# 1  0.596    0.595 0.0728   928.5.67e-126     1    760 -1514. -1501.
# # 3 more variables: deviance <dbl>, df.residual <int>, nobs <int>
```

- `augment()` returns a data frame of residuals and predictions (fitted values) for the model realisation.

```
augment(model)
# # A tibble: 632 × 8
# leave  degree .fitted   .resid  .hat .sigma  .cooksd .std.resid
# <dbl>  <dbl>  <dbl>     <dbl>   <dbl> <dbl>   <dbl>     <dbl>
# 1 0.579 -0.211    0.539  0.0398  0.00165 0.0728 0.000247   0.547
# 2 0.678 -0.748    0.587  0.0914  0.00247 0.0728 0.00195    1.26
# 3 0.386  1.63     0.376  0.00957 0.00582 0.0729 0.0000509  0.132
# 4 0.653 -0.964    0.606  0.0473  0.00306 0.0728 0.000648   0.650
# ...
```

The advantage of generating model diagnostics and outputs that are tidy is that it eases the process of working with many model realisations. This is a common requirement for modern data analysis, where statistical inferences are made empirically from resampling. For example, we may wish to generate single-variable linear regression models separately for each selected explanatory variable. We could use these outputs to annotate the scatterplots in Figure 6.2 by their regression line and colour observations according to their residual values, distance from the regression line. These models can be generated with reasonably little code by making use of the package `broom` and a style of functional programming in R, which is supported by the `purrr` package.

Example code:

```
single_model_fits <- cons_data |>
  mutate(across(c(younger:heavy_industry), ~(.x-mean(.x))/sd(.x))) |>
  pivot_longer(
    cols=younger:heavy_industry,
    names_to="expl_var", values_to="z_score"
  ) |>
  # Nest to generate list-column by expl_var.
  nest(data=-expl_var) |>
  mutate(
    # Use map() to iterate over the list of datasets.
    model = map(data, ~lm(leave ~ z_score, data = .x)),
    # glance() for each model fit.
    fits = map(model, glance),
    # tidy() for coefficients.
    coefs = map(model, tidy),
    # augment() for predictions/residuals.
    values=map(model, augment),
  )


single_model_fits |>
  # Unnest output from glance.
```

```
unnest(cols = fits) |>
# Remove other list-columns.
select(-c(data, model))
```

```
# # A tibble: 10 × 15
# expl_var   r.squared adj.r.sq  sigma  stat    p.value  df logLik    AIC
# <chr>           <dbl>    <dbl>  <dbl> <dbl>      <dbl> <dbl> <dbl> <dbl>
# 1 younger       0.289    0.288 0.0965  257. 1.05e- 48   1   582. -1158.
# 2 own_home      0.185    0.184 0.103   143. 7.42e- 30   1   539. -1071.
# 3 no_car        0.157    0.155 0.105   117. 3.81e- 25   1   528. -1050.
# 4 white         0.169    0.168 0.104   128. 3.79e- 27   1   532. -1059.
# 5 eu_born       0.233    0.232 0.100   191. 3.42e- 38   1   558. -1110.
# 6 christian     0.238    0.236 0.100   196. 4.95e- 39   1   560. -1114.
# 7 professi...   0.320    0.319 0.0944  296. 1.08e- 54   1   596. -1186.
# 8 degree        0.596    0.595 0.0728  928. 5.67e-126   1   760. -1514.
# 9 not_good...   0.316    0.315 0.0947  291. 5.93e- 54   1   594. -1182.
# 10 heavy_in...  0.504    0.503 0.0806  640. 5.43e- 98   1   696. -1385.
# #  6 more variables: BIC <dbl>, deviance <dbl>, df.residual <int>,
# #    nobs <int>, coefs <list>, values <list>
```

Code description:

1. *Setup*: In order to generate separate models for separate explanatory variables, we need to generate nested data frames. These are data frames stored in a special type of column (a list-column) in which the values of the column is a list of data frames – one for each explanatory variable over which we would like to compute a model. You can think of parameterising nest() in a similar way to group_by(). We first pivot_longer() to generate a data frame where each observation contains the recorded Leave vote for a constituency and its corresponding z_score value for each explanatory variable. There are 10 explanatory variables and so nest() returns a data frame with the dimensions 10x2 – a variable identifying the explanatory variable on which the model is to be built (expl_var) and a list-column, each element containing a data frame with the dimensions 632x13.

2. *Build model*: In mutate(), purrr's map() function is used to iterate over the list of datasets and fit a model to each nested dataset. The new column model is a list-column this time containing a list of model objects.

3. *Generate outputs*: Next, the different cuts of model outputs can be made using glance(), tidy(), augment(), with map() to iterate over the list of model objects. The new columns are now list-columns of data frames containing model outputs.

4. *Extract outputs*: Finally we want to extract the values from these

nested data. This can be achieved using `unnest()` and supplying to
the `cols` argument the names of the `list-columns` from which we want
to extract values.

### 6.3.3 Plot models tidily

In Figure 6.4 estimated regression coefficients are plotted from a multivariate
model, annotated with 95% Confidence Intervals. The ggplot2 specification is
reasonably straightforward.

The code for Figure 6.4:

```
model <- cons_data |>
  mutate(across(c(younger:heavy_industry), ~(.x-mean(.x))/sd(.x))) %>%
  lm(leave ~ degree + eu_born + white + no_car + not_good_health +
    heavy_industry, data=.)

tidy(model) |>
  filter(term != "(Intercept)") |>
  ggplot(
    aes(x=reorder(term, -estimate),
        y=estimate, ymin=estimate-1.96*std.error,
        ymax=estimate+1.96*std.error)
        ) +
  geom_pointrange() +
  coord_flip()
```

The plot specification:

1. *Data*: A data frame of model coefficients extracted from the multivariate model object (`model`) using `tidy()`.
2. *Encoding*: y-position varies according to the size of the coefficient
   `estimate` and the 95% confidence intervals, derived from `std.error`
   and encoded using `ymin` and `ymax` parameters.
3. *Marks*: `geom_pointrange()`, which understands `ymin` and `ymax`, for the
   dots with confidence intervals.
4. *Setting*: `coord_flip()` to make variable names easier to read.

### 6.3.4 Extend model terms

To include a Fixed Effect (FE) term on region, the `region` variable is simply
added as a variable to `lm()`. However, we must convert it to a factor variable; this
has the effect of creating dummies on each value of `region`. Default behaviour
within `lm()` is to hold back a reference value of region with FE regression
coefficients describing the effect on the outcome of a constituency located

in a given region relative to that reference region. So the reference region (intercept) in the model below is East Midlands – the first in the factor to appear alphabetically. The signed coefficient estimates for regions identifies whether, after controlling for variation in demographics, the Leave vote for a particular region is expected to be higher or lower than this.

```
cons_data |>
  mutate(
    across(c(younger:heavy_industry), ~(.x-mean(.x))/sd(.x)),
    region=factor(region)) %>%
  lm(leave ~ region + degree + eu_born + white  + no_car +
    not_good_health + heavy_industry, data=.) |>
  tidy()
```

```
# # A tibble: 17 × 5
# term                     estimate std.error statistic  p.value
# <chr>                       <dbl>     <dbl>     <dbl>    <dbl>
#  1 (Intercept)             0.530    0.00581     91.3   0
# 2 regionEast of England    0.00363  0.00787      0.462 6.45e- 1
# 3 regionLondon             0.0654   0.00948      6.90  1.30e-11
# 4 regionNorth East         0.00482  0.00945      0.510 6.10e- 1
# 5 regionNorth West        -0.0200   0.00728     -2.75  6.12e- 3
# 6 regionScotland          -0.145    0.00843    -17.2   1.28e-54
# 7 regionSouth East         0.00377  0.00752      0.502 6.16e- 1
# 8 regionSouth West        -0.0233   0.00789     -2.95  3.26e- 3
# 9 regionWales             -0.0547   0.00860     -6.36  3.87e-10
# 10 regionWest Midlands     0.0236   0.00745      3.17  1.59e- 3
# 11 regionYorkshire         0.0112   0.00762      1.47  1.41e- 1
# 12 degree                 -0.0772   0.00339    -22.8   1.30e-83
# 13 eu_born                 0.0163   0.00308      5.29  1.72e- 7
# 14 white                   0.0303   0.00314      9.66  1.18e-20
# 15 no_car                 -0.0336   0.00292    -11.5   6.50e-28
# 16 not_good_health         0.0102   0.00331      3.07  2.24e- 3
# 17 heavy_industry          0.0132   0.00266      4.96  9.23e- 7
```

We want our model to represent a dummy for every region, and so we add -1 to the specification. Doing this removes the intercept or reference region, making $R^2$ no longer meaningful.

```
cons_data |>
  mutate(
    across(c(younger:heavy_industry), ~(.x-mean(.x))/sd(.x)),
    region=factor(region)) %>%
  lm(leave ~ region + degree + eu_born + white  + no_car +
```

```
      not_good_health + heavy_industry -1, data=.) |>
  glance()
```

```
# # A tibble: 1 × 12
# r.squared adj.r.squared sigma statistic p.value df logLik AIC
# <dbl>   <dbl>       <dbl>  <dbl> <dbl>  <dbl> <dbl>  <dbl> <dbl>
#   1     0.995   0.995 0.0371  7625.    0    17 1193. -2351.
# #  3 more variables: deviance <dbl>, df.residual <int>, nobs <int>
```

To include an Interaction on region, we need to set a variable that will be used to represent these regional constants (`cons`), and the Interaction is added with the notation `:`.

```
model <- cons_data |>
  mutate(
    across(c(younger:heavy_industry), ~(.x-mean(.x))/sd(.x)),
    region=as.factor(region), cons=1) %>%
  lm(leave ~ 0 +
      (cons + degree  + eu_born + white  + no_car + not_good_health +
       heavy_industry):(region),
    data=.
    )
```

The model updated with the regional Interaction term results in many more coefficients that are, as discussed, somewhat unstable. To plot them, as in Figure 6.7, we minimally update the code used to generate the previous model outputs.

```
tidy(model) |>
  separate(term, into= c("term", "region"), sep=":") |>
  mutate(region=str_remove(region,"region")) |>
  filter(term!="cons") |>
  ggplot() +
  geom_col(aes(x=reorder(term, -estimate), y=estimate), alpha=.3)+
  geom_pointrange(aes(
    x=reorder(term, -estimate),y=estimate,
    ymin=estimate-1.96*std.error, ymax=estimate+1.96*std.error
  )) +
  geom_hline(yintercept = 0, size=.2)+
  facet_wrap(~region) +
  coord_flip()
```

The plot specification:

1. *Data*: A data frame of model coefficients extracted from the multi-variate model object using `tidy()`. To make clean plot labels we need to remove unnecessary text in the `term` variable (e.g. "cons:regionEast Midlands"). `separate()` allows us to split this column on : and then `str_remove()` is quite obvious. We do not wish to plot the FE constants and so `filter()` them out.

2. *Encoding*: y-position varies according to the size of the coefficient estimate and the 95% confidence intervals, in exactly the same way as for Figure 6.4.

3. *Marks*: `geom_pointrange()`, encoded as in Figure 6.4. The only addition is light bars in the background (`geom_col()`). This seems to aid interpretation of the direction and size of the coefficients.

4. *Facets*: `facet_wrap()` on `region` in order to display coefficients estimated separately for each region.

### 6.3.5　Evaluate models with lineups

In Figure 6.8 is a map line-up of the residuals from FE-updated model – our expectation is that these residuals should *no longer* be spatially autocorrelated, since we collapse regional varation into our FE term.

Using functional-style programming, and index{packages!tidymodels} tidymodels, plot lineups can be generated with surprisingly paired-back code. First generate a model object and extract residuals from it, again making use of `nest()`, `map()` and `augment()`:

```
model <- cons_data |>
  select(-c(population, population_density)) |>
  mutate(
    across(c(younger:heavy_industry), ~(.x-mean(.x))/sd(.x)),
    type="full_dataset", region=as.factor(region)
    ) |>
  nest(data=-type) |>
  mutate(
    # Include `-1` to eliminate the constant term and include
    # a dummy for every area.
    model=map(data,
      ~lm(leave ~ region +  degree  + eu_born + white  + no_car +
      not_good_health + heavy_industry -1, data=.x)
      ),
    # augment() for predictions / residuals.
    values=map(model, augment)
  )
```

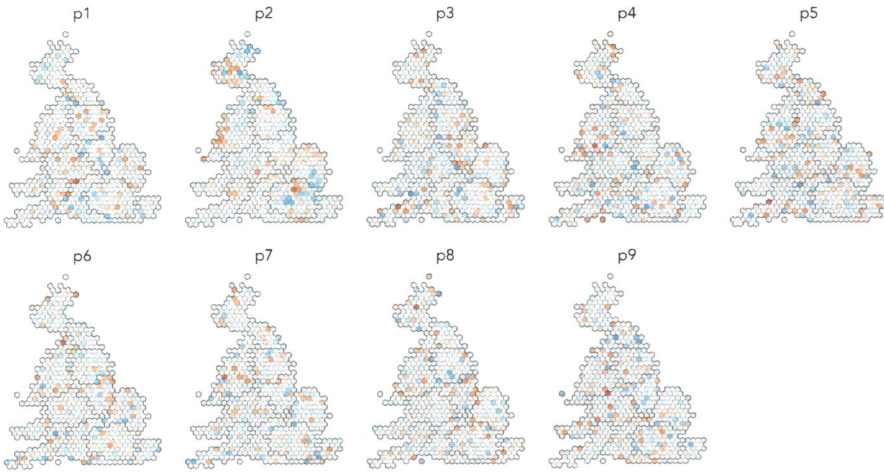

**Figure 6.8:** Map line-up of residuals from model with Fixed Effect on region. The 'real' dataset is presented alongside 8 decoy plots generated by randomly permuting the observed residuals around constituencies. Adding the FE term has addressed some of the systematic over- and under-estimation of the vote between regions (compare for example Figure 6.5). There is nevertheless obvious spatial autocorrelation, plot 2 being the real data. Further analysis, for example of the constituencies for which Leave is particularly over- and under-represented, may be instructive.

Next, generate permuted data by randomly shuffling residual values around constituencies. To do this requires some knowledge of the `rsample` package and its functions. We extract residuals from the `list-column` named `values`, remove redundant `list-columns` (with `select()`) and then `unnest()` on the original data and the new `resids` field to return to a dataset where each row is a constituency, but now containing residual values for the multivariate model. From here, we use the `permutations()` function from `rsample` to shuffle the constituency ID column (`pcon19cd`) randomly around, generating eight permuted datasets and appending the real data (`apparent=TRUE`). This results in a new data frame where each row contains a permuted dataset, stored in a `list-column` named `splits` and labelled via an `id` column. We need to `map()` over `splits` to convert each split object into a data frame, using `rsample`'s `analysis()` function. From here, we `unnest()` to generate a dataset where each row is a constituency and its corresponding residual value (real or shuffled) for a given permutation `id`.

```
permuted_data <- model |>
  mutate(
    resids=map(values, ~.x |> select(.resid))
  ) |>
  select(-c(model, values)) |>
  unnest(cols=c(data,resids)) |>
  select(pcon19cd, .resid) |>
  permutations(permute=c(pcon19cd), times=8, apparent=TRUE) |>
  mutate(data=map(splits, ~rsample::analysis(.))) |>
  select(id, data) |>
  unnest(cols=data)
```

Now that we have the permuted dataset, the lineup can be generated straight-forwardly with standard ggplot2 code:

```
# Store max value of residuals for setting limits in map colour scheme.
max_resid <- max(abs(permuted_data$.resid))
# Store vector of permutation IDs for shuffling facets in the plots.
ids <- permuted_data |> pull(id) |> unique()

cons_hex |>
  select(cons_code, region) |>
  inner_join(permuted_data, by=c("cons_code"="pcon19cd")) |>
  mutate(id=factor(id, levels=sample(ids))) |>
  ggplot() +
  geom_sf(aes(fill=.resid), colour="#636363", linewidth=0.05)+
  geom_sf(
    data=. %>% group_by(region) %>% summarise(),
    colour="#636363", linewidth=0.2, fill="transparent"
  )+
  facet_wrap(~id, ncol=3) +
  scale_fill_distiller(palette="RdBu", direction=1,
                       limits=c(-max_resid, max_resid), guide="none")
```

The plot specification:

1. *Data*: `inner_join` the permuted data on the simple features file containing the hexagon cartogram boundaries (`cons_hex`). To generate the lineup we `facet_wrap()` on the permutation `id`. By default ggplot2 will draw facets in a particular order – determined either by the numeric or alphabetical order of the facet variable's values, or by an order determined by a factor variable. Each time we plot the lineups, we want the order in which the real and decoy plots are drawn to vary. Therefore we convert `id` to a factor variable and shuffle the

levels (the ordering) around, using the `sample()` function on a vector of permutation IDs (`ids`) before piping to `ggplot()`. Note that we also record the maximum absolute value of the residuals to ensure that they are coloured symmetrically on 0 (`max_resid`). Finally, you may notice there are two `geom_sf()` calls in the plot specification. The second draws regional boundary outlines across each plot. This is achieved by collapsing the hexagon data on region (using `group_by()` and `summarise()`).

2. *Encoding*: hexagons are filled according to the residual values (`fill=.resid`).

3. *Marks*: `geom_sf()` for drawing the hexagon outlines. The first `geom_sf` colours each constituency on its residual value. The second does not encode any data values – notice there is no `aes()` – and is simply used to draw the region outlines.

4. *Facets*: `facet_wrap()` on `region` in order to display coefficients estimated separately for each region.

5. *Scale*: `scale_fill_distiller()` for ColorBrewer (Harrower and Brewer 2003) scheme, using the RdBu palette and with `limits` set to `max_resid`.

6. *Setting*: The `linewidth` parameter of the hexagon outlines is varied so that the regional outlines in the second call to `geom_sf()` appear more salient. Also here, a transparent `fill` to ensure that the regional outlines do not occlude the encoded residuals.

## 6.4 Conclusion

This chapter demonstrated how visual and computational approaches can be used together in a somewhat more 'traditional' area-level regression analysis. Associations between constituency-level Leave voting in the UK's 2016 EU Referendum and selected variables describing the demographic and socio-economic composition of constituencies were explored, with data graphics used to characterise bias in the generated models – to identify geographic and regional groupings that our early models ignore. Two classes of model update for addressing this geographic grouping were covered: those that treat geographic dependence in the values of variables as a nuisance term that is to be quantified and controlled away, and those that explicitly try to model for geographic grouping in processes. We introduced some initial techniques for dealing with both, treating geography as a categorical variable: a Fixed Effect term to assess regional dependence and Interaction term to assess regional non-stationarity. Importantly, the chapter reused some of the `dplyr` and functional programming code templates instrumental for working over models. There was a step-up in code complexity. Hopefully you will see in the next chapter that this sort of functional programming style (Wickham, Çetinkaya-Rundel, and Grolemund 2023) greatly aids the process of performing and working with resampled datasets, a key feature of modern computational data analysis.

## 6.5   Further Reading

An area-level analysis of the Brexit vote:

- Beecham, R., Williams, N. and Comber, L. 2020. "Regionally-structured explanations behind area-level populism: An update to recent ecological analyses." *PLOS One*, 15(3): e0229974. doi: 10.1371/journal.pone.0229974.

On modelling for geographic dependence and non-stationarity:

- Comber, A., Brunsdon, C., Charlton, M. et al. 2023. "A route map for successful applications of Geographically Weighted Regression." *Geographical Analysis*, 55 (1): 155–178. doi: 10.1111/gean.12316.
- Wolf, L. J. et al., 2023. "On Spatial and Platial Dependence: Examining Shrinkage in Spatially Dependent Multilevel Models." *Annals of the American Association of Geographers*, 55(1): 1–13. doi: 10.1080/24694452.2020.1841602.

The original graphical inference paper:

- Buja, A., Cook, D., Hofmann, H., Lawrence, M., Lee, E.K., Swayne, D. F. and Wickham, H. 2010. "Statistical Inference for Exploratory Data Analysis and Model Diagnostics." *Royal Society Philosophical Transactions A*, 367:4361–83. doi: 10.1098/rsta.2009.0120.

A guide to model building in the `tidyverse`:

- Ismay, C. and Kim, A. 2020. "Statistical Inference via Data Science: A ModernDive into r and the Tidyverse", New York, NY: *CRC Press*. doi: 10.1201/9780367409913.
- Kuhn, M. and Silge, J. 2023. "Tidy Modelling with R.", Sebastopol, CA: *O'Reilly*.

A quick guide to functional programming in R and `tidyverse`:

- Wickham, H., Çetinkaya-Rundel, M., Grolemund, G. 2023, "R for Data Science, 2nd Edition", Sebastopol, CA: *O'Reilly*.
  – Chapter 25, 26.

# 7

## *Uncertainty*

By the end of this chapter you should gain the following knowledge and practical skills.

---

**ℹ Knowledge**

☐ Appreciate the main challenges and objectives of uncertainty representation.

☐ Learn how visualization techniques can be used to support 'frequency framing'.

☐ Understand how parameter uncertainty due to random fluctuation can be estimated computationally.

---

**ℹ Practical skills**

☐ Generate estimates of parameter uncertainty using bootstrap resampling.

☐ Apply functional-style programming for working over bootstrap resamples.

☐ Write ggplot2 code to create uncertainty visualizations: icon arrays, risk theatres, gradient bars, ensemble and hypothetical outcome plots.

---

## 7.1 Introduction

Uncertainty is a key preoccupation of those working in statistics and data analysis. A lot of time is spent providing estimates for it, reasoning about it and trying to take it into account when making evidence-based claims and decisions. There are many ways in which uncertainty can enter a data analysis and many ways in which it can be conceptually represented. This chapter focuses mainly on parameter uncertainty: quantifying and conveying the different possible values that a quantity of interest might take. It is straightforward to imagine how visualization can support this. We can use data graphics to represent different values and give greater emphasis to those for which we have more certainty – to communicate or imply levels of uncertainty in the background. Such representations are nevertheless quite challenging to execute. In Chapter 3 we learnt that there is often a gap between the visual encoding of data and its perception. There is a tendency in standard data graphics to imbue data with marks that over-imply precision. We will consider research in Cartography and Information Visualization on uncertainty representation, before exploring and applying techniques for visually encoding parameter uncertainty. We will do so using STATS19 road safety data, exploring how injury severity rates in pedestrian-vehicle crashes vary over time and by geographic area.

## 7.2   Concepts

### 7.2.1   Uncertainty visualization

Cartographers and Information Visualization researchers have been concerned for some time with *visual variables*, or *visual channels* (Munzner 2014), that might be used to encode uncertainty information. Figure 7.1 displays several of these. Ideally, visual variables should be intuitive, logically related to notions of precision and accuracy, while also allowing sufficient discriminative power when deployed in data dense visualizations.

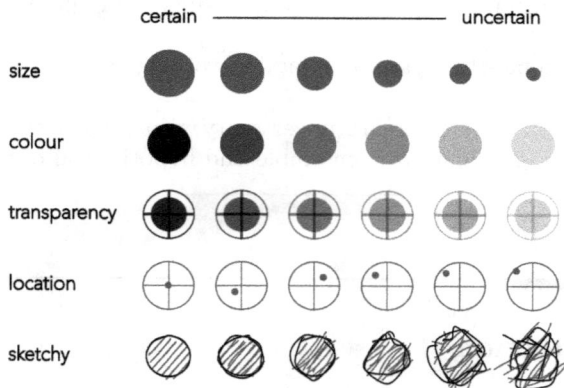

**Figure 7.1:** Visual variables that can be used to represent levels of uncertainty information. Sketchy rendering is generated with `Rough.js`, an implementation of the work published in Wood et al. (2012).

Kinkeldey, MacEachren, and Schiewe (2014) provides an overview of empirical research into the effectiveness of proposed visual variables against these criteria. As intuitive signifiers of uncertainty, or lack of precision, *fuzziness* (not encoded in Figure 7.1) and *location* have been shown to work well. Slightly less intuitive, but nevertheless successful in terms of discrimination, are *size*, *transparency* and *colour value*. *Sketchiness* is another intuitive signifier proposed in Boukhelifa et al. (2012). As with many visual variables, sketchiness is probably best considered as an ordinal visual variable to the extent that there is a limited range of sketchiness levels that can be discriminated. An additional feature of sketchiness is its sense of informality. This may be desirable in certain contexts, less so in others (see Wood et al. 2012 for further discussion).

When thinking about uncertainty visualization, a key guideline is that:

---

"Things that are not precise should not be encoded with symbols that look precise."

---

Much discussed in recent literature on uncertainty visualization (e.g. Padilla, Kay, and Hullman 2021) is the US National Weather Service's (NWS) (NHC 2023) cone graphic (Figure 7.2). The cone starts at the storm's current location and spreads out to represent the modelled *projected path* of the storm. The main problem is that the cone implies the storm is expanding as it moves away from its current location, when this is not the case. In fact there is more uncertainty in the areas that could be affected by the storm the further away those areas are from the storm's current location. The second problem is that the cone uses strong lines that imply precision. The temptation is to think that anything contained by the cone is unsafe and anything outside of it is safe. This is of course not what is suggested by the model. Rather, that areas not contained by the cone are beyond some chosen threshold probability. You will notice that the graphic in Figure 7.2 is annotated with a guidance note to discourage such false interpretation.

In Van Goethem et al.'s (2014) redesign, colour value is used to represent four binned categories of storm probability suggested by the model. Greater visual saliency is therefore conferred to locations where there is greater certainty. The state boundaries are also encoded somewhat differently. In Figure 7.3 US states are symbolised using a single line generated via curve schematisation (Van Goethem et al. 2014). The thinking here is that hard lines in maps tend to induce binary judgements. If the cone is close to but not overlapping a state boundary, for example, should a state's authorities prepare and organise a response any differently from a state whose boundary very slightly overlaps the cone? Context around states is therefore introduced in the redesign, but in a way that discourages binary thinking; precise inferences of location are not possible as the state areas and borders are very obviously not exact.

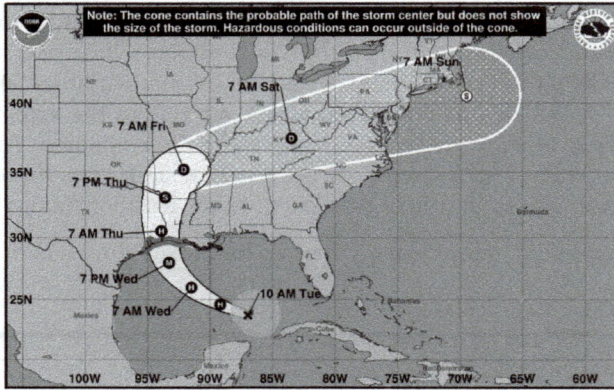

**Figure 7.2:** National Hurricane Center cone design showing probable track of the centre of a cyclone. Cone graphic re-printed with permission, credit National Oceanic and Atmospheric Administration/National Weather Service.

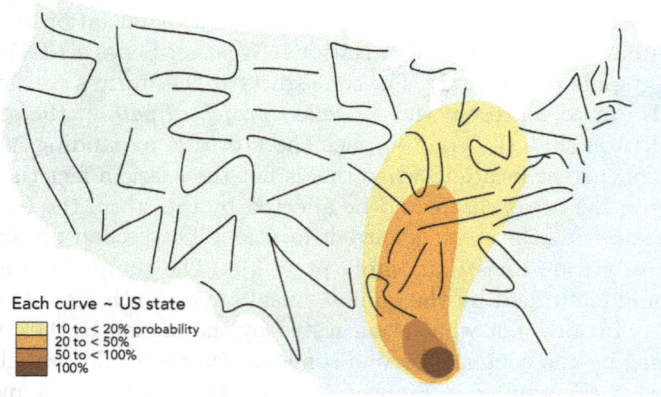

**Figure 7.3:** Permitted author edited reproduction of Van Goethem et al.'s (2014) design for probable cyclone track.

## 7.2.2   Frequency framing

For practical reasons the rest of the chapter considers how these general principles for uncertainty representation might be applied to a single aspect of uncertainty: quantifiable parameter uncertainty. Parameters of interest are often probabilities or relative frequencies – ratios and percentages describing the probability of some event happening. It is notoriously difficult to develop intuition around these sorts of relative frequencies, and so data graphics can usefully support their interpretation.

In our STATS19 road crash dataset, a parameter of interest is the pedestrian injury severity rate, or the proportion of all pedestrian crashes that result in serious or fatal injury (KSI). We might wish to compare the injury severity rate of crashes taking place between two local authority areas, say Bristol and Sheffield. There is in fact quite a difference in the injury severity rate between these two local authorities. In 2019, 35 out of 228 reported crashes (15%) in Bristol were KSI, while for Sheffield this figure was 124 out of 248 reported crashes (50%). This feels like quite a large difference, but it is difficult to imagine or experience these differences in probabilities when written down or encoded visually using relative bar length in standardised bar charts.

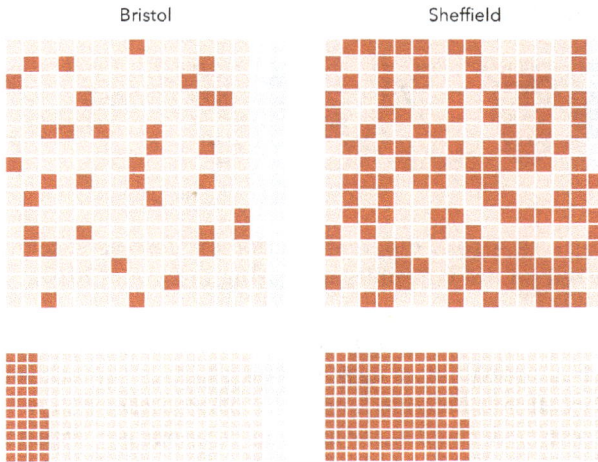

**Figure 7.4:** Icon array displaying injury severity rates for Pedestrian-Vehicle crashes.

Icon arrays are used in public health communication and have been demonstrated to be effective at communicating probabilities of event outcomes. They offload the thinking that happens when evaluating ratios. The icon arrays in Figure 7.4 communicate the two injury severity rates for Bristol and Sheffield. Each crash is a square, and crashes are coloured according to whether they resulted in a serious injury or fatality (dark red) or slight injury (light red). In the bottom row, cells are given a non-random ordering to effect something similar to a standardised bar chart. While the standardised bars enables the two recorded proportions to be "read-off" (15% and 50% KSI), the random arrangement of cells in the icon array perhaps builds intuition around the differences in *probabilities* of a pedestrian crash resulting in serious injury.

There are compelling examples of icon arrays being used in data journalism, most obviously to communicate outcome probabilities in political polling. You might remember that at the time of the 2016 US Presidential election there was much criticism levelled at pollsters, even the excellent FiveThirtyEight

(Silver 2016), for not correctly calling the result. Huffpost gave Trump a 2% chance of winning the election, The New York Times 15% and FiveThirtyEight 28%. Clearly the Huffpost estimate was really quite off, but thinking about FiveThirtyEight's prediction, how surprised should we be if an outcome that is predicted to happen with a probability of almost a third, does in fact occur?

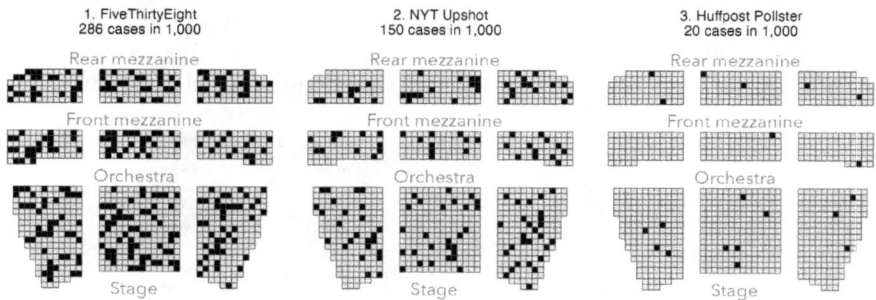

**Figure 7.5:** Risk theatre of different election eve forecasts, reimplemented in ggplot2 but based on data graphics appearing in Gross (2016).

The risk theatre (Figure 7.5) is a variant of an icon array. In this case it represents polling probabilities as seats of a theatre – a dark seat represents a Trump victory. If you imagine buying a theatre ticket and being randomly allocated to a seat, how confident would you be about not sitting in a "Trump" seat in the FiveThirtyEight image? The distribution of dark seats suggests that the 28% risk of a Trump victory according to the model is not negligible.

### 7.2.3   Quantifying uncertainty in frequencies

In the icon arrays above we made little of the fact that the sample size varies between the two recorded crash rates. This was because the differences were in fact reasonably small. When looking at injury severity rates across all local authorities in the country, however, there is substantial variation in the rates *and* sample sizes. Bromsgrove has a very low injury severity rate based on a small sample size (4%, or one out of 27 crashes resulting in KSI); Cotswold has a very high injury severity rate based on a small sample size (75%, or 14 out of 19 crashes resulting in KSI). With some prior knowledge of these areas one might expect the difference in KSI rates to be in this direction, but would we expect the difference to be of this order of magnitude? Just three more KSIs recorded in Bromsgrove takes its KSI rate up to that of Bristol's.

Although STATS19 is a population dataset to the extent that it contains data on every crash recorded by the police, it makes sense that the more data on which our KSI rates are based, the more certainty we have in them being reliable estimates of injury severity – ones that might be used to predict injury severity in future years. So we can treat our observed injury severity (KSI) rates as being derived from samples of an (unobtainable) population. Our calculated KSI rates are *parameters* that try to represent, or estimate, this population.

Although this formulation might seem unnecessary, from here we can apply some statistical concepts to quantify uncertainty around our KSI rates. We assume:

1. The variable of interest, KSI rate, has an unobtainable *population* mean and standard deviation.
2. That our data are one *sample* from this unobtainable population, but other samples could be drawn that will result in different outcomes, estimated KSI rates, simply by chance.
3. From any *sample* that is drawn we can calculate a mean and standard deviation in KSI rates.
4. And so we can derive a *sampling distribution* and obtain an array of estimated KSI rates and other parameters from *re*sampling many times.
5. This *sampling distribution* could then be used to quantify how precise are our estimates of KSI rate. Generally the larger the sampling distribution, the more precise, the less uncertain, the estimate.

In Chapter 6 we used Confidence Intervals to estimate the uncertainty around regression coefficients. From early stats courses you might have learnt how Confidence Intervals can be calculated using statistical theory, but we can derive them empirically via bootstrapping – the process enumerated above. So a bootstrap resample involves taking a random sample with replacement from the original data and of the same size as the original data. From this resample a parameter estimate can be derived, in this case the KSI rate. And this process can be repeated many times to generate an empirical *sampling distribution* for the parameter. The standard error can be calculated from the standard deviation of the sampling distribution. This non-parametric bootstrapping approach is especially useful in exploratory analysis (Beecham and Lovelace 2023): it can be applied to many sample statistics, makes no distributional assumptions and can work on quite complicated sampling designs.

Presented in Figure 7.6 are KSI rates with error bars used to display 95% Confidence Intervals generated from a bootstrap procedure in which 1000 resamples were taken with replacement. Upper and lower limits were lifted from .025 and .975 percentile positions of the bootstrap sampling distribution. Assuming that the observed data are drawn from a wider (unobtainable) population, the 95% Confidence Intervals demonstrate that while Cotswold recorded a very large KSI rate, sampling variation means that this figure could be much lower (or higher), whereas for Bristol and Sheffield, where our KSI rate is derived from more data, the range of plausible values that the KSI rate might take due to sampling variation is much smaller – there is less uncertainty associated with their KSI rates.

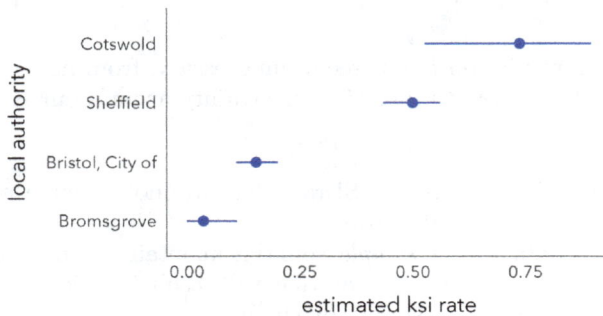

**Figure 7.6:** KSI rates for pedestrian-vehicle crashes in selected local authorities with bootstrapped CIs (derived from 1000 resamples).

### 7.2.4   Visualizing uncertainty in frequencies

Error bars, like those in Figure 7.6, are a space-efficient way of conveying parameter uncertainty. However, remembering our main guideline for uncertainty visualization – that *things that are not precise should not be encoded with symbols that look precise* – they do have problems. The hard borders can lead to binary or categorical thinking (see Correll and Gleicher 2014). Certain values within a Confidence Interval are more probable than others, and so we should endeavour to use a visual encoding that reflects this. Matt Kay's excellent ggdist package (Kay 2024) extends ggplot2 with a range of chart types for representing these sorts of intervals. In Figure 7.7 error bars are replaced by *half eye plots* and *gradient bars*, which give greater visual saliency to values of KSI that are more likely.

STATS19 road crash data are released annually. Given the wide uncertainty bands for some local authorities, it might be instructive to explore the stability of KSI rates year-on-year. In Figure 7.6 these KSI rates are represented with a bold line, and the faint lines are superimposed bootstrap resamples. The lines demonstrate volatility in the KSI rates for Cotswold and Bromsgrove due to small numbers. The observed increase in KSI rates for Sheffield since 2015 does appear to be a genuine one, although may also be affected by uncertainty around data collection and how reliably injury severity is recorded in the dataset.

The superimposed lines in the figure above are a form of ensemble visualization. An alternative approach might have been to animate over the bootstrap resamples to generate a Hypothetical Outcome Plot (HOP) (Hullman, Resnick,

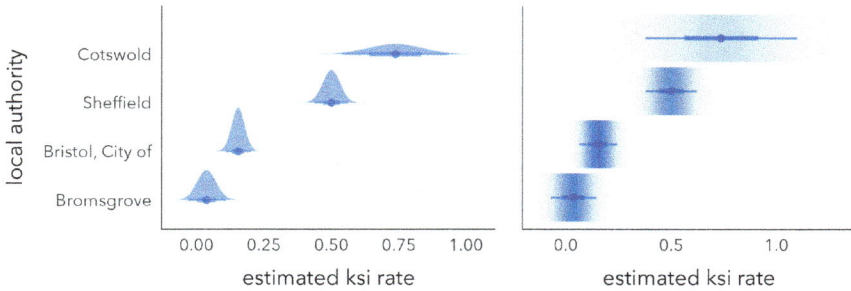

**Figure 7.7:** KSI rates for pedestrian-vehicle crashes in selected local authorities, with bootstrapped uncertainty estimates.

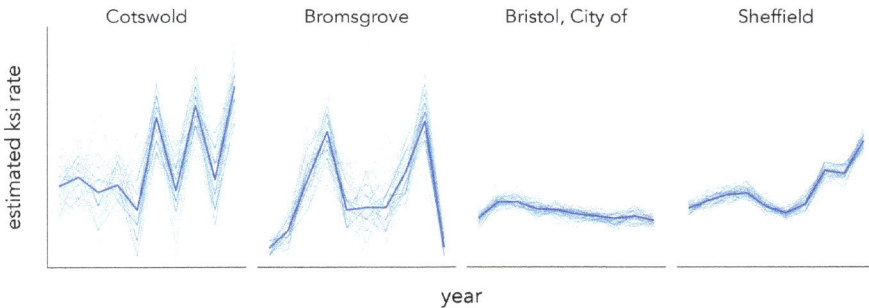

**Figure 7.8:** Year-on-year KSI rates for pedestrian-vehicle crashes in selected local authorities, with bootstrap resamples superimposed.

and Adar 2015). HOPs convey a sense of uncertainty by animating over random draws of a distribution. As there is no single outcome to anchor to, HOPs force viewers to account for uncertainty, recognising that some less probable outcomes may also be possible – essentially to think distributionally.

### 7.2.5 Multiple comparisons

In road safety monitoring, a common ambition is to compare crash rates across local authorities. This is in order to make inferences around patterns of high and low injury severity rate. We might represent injury severity rates as Risk Ratios (RR) comparing the observed injury severity rate in each local authority to a benchmark, say the injury severity rate we would expect to see nationally. RRs are an intuitive measure of effect size: RRs >1.0 indicate that the injury severity rate is greater than the national average; RRs <1.0 that it is less than the national average. As they are a ratio of ratios, and therefore agnostic to

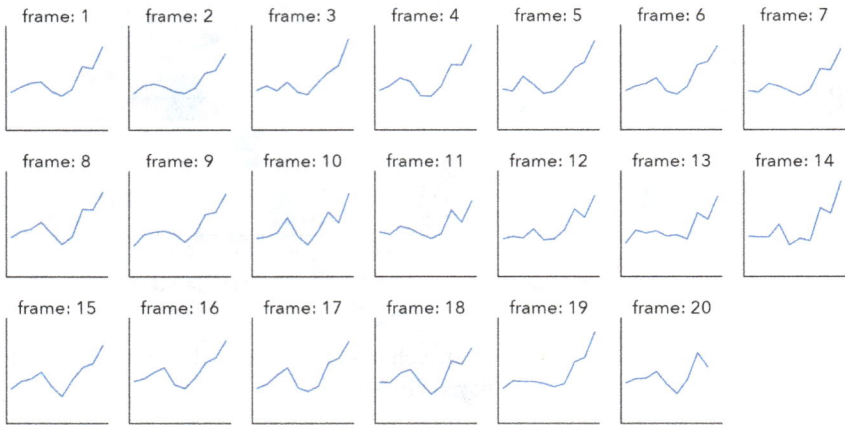

**Figure 7.9:** Frames from hypothetical outcome plot of year-on-year KSI rates for pedestrian-vehicle crashes.

sample size, RRs can nevertheless be unreliable. Two ratios might be compared that have very different sample sizes, and no compensation is made for the one that contains more data.

We can use quantitative measures to adjust for this. In the example in Figure 7.10 we use hierarchical modelling to shrink local authority KSI rates towards the global mean (national average KSI rate) where they are based on small numbers of observations (see Beecham and Lovelace 2023). From here our effect sizes, called Bayesian Risk Ratios, are sensitive to uncertainty since they are made more conservative where they are based on fewer observations. The Bayesian Risk Ratio for each local authority is represented with a | icon: angled to the right / where the KSI rate is greater than expected, to the left \ where it is less than expected. Additionally, we use bootstrap resampling to derive confidence intervals for our Bayesian RRs. If this interval does not cross 1.0, the RR is judged statistically significant and is coloured according to whether estimated RRs are above (/) or below (\) expectation.

From Figure 7.10 we make inferences around concentrations of high and low injury severity rate (in the annotations). A problem with this approach, and explicitly encoding 'statistical significance' values, is one familiar to statisticians but that is rarely addressed in visual data analysis: the multiple comparison problem. Whenever a statistical signal is identified, there is a chance that the result observed is in fact a false alarm. In the plot above which uses a 95% confidence level, the "false positive rate" is expected to be 5% or 1/20. When many tests are considered simultaneously, as in Figure 7.10, the number of these false alarms begins to accumulate. There are corrections that can be used to address this: test statistics can be adjusted and made more

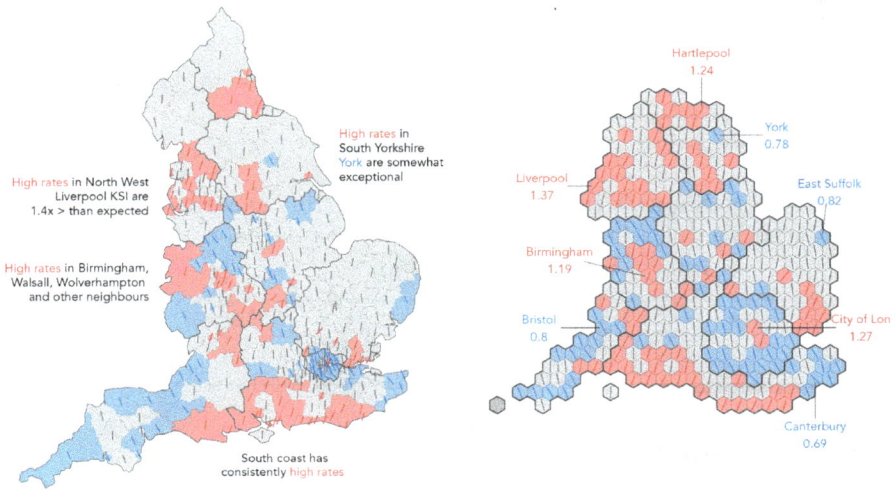

**Figure 7.10:** Bayesian Risk Ratios comparing pedestrian injury severity rates in English local authorities coloured according to 'statistical significance', whether the bootstrap confidence interval does not cross 1.0.

conservative. But these corrections have consequences. Too severe a correction can result in statistical tests that are underpowered and result in an elevated false negative rate, where a statistical test fails to detect an effect that truly exists. See Brunsdon and Charlton (2011) for an interesting discussion in the context of mapping crime rates.

So there is no single solution to multiple testing, which happens often in visual data analysis, especially in Geography, where health and other outcomes are mapped visually. It is actually less of problem in Figure 7.10 since our RRs are derived from a multilevel model in which estimates are partially pooled, or shrunk, to reflect the level of information we have (Gelman, Hill, and Yajima 2012). Presenting the RRs in their spatial context, and providing full information around RRs that are not significant (the oriented lines), also supports informal calibration. For example, depending on the phenomena, we may wish to attach more certainty to RRs that are labelled statistically significant and whose direction is consistent with their neighbours than those that are exceptional from their neighbours. Additionally, constructing a graphical line-up test (Wickham et al. 2010) allows us to explore whether the sorts of spatial patterns in RR values in the observed data are genuine or might appear in random decoy maps. Although informal, this sort of visual test approximates to the type of question that transport analysts may ask when identifying priority areas for road safety intervention (Beecham and Lovelace 2023).

> **Task**
>
> Watch Matt Kay's excellent talk to BostonCHI, *Uncertainty Visualization as a Moral Imperative*:
>
> - https://www.youtube.com/watch?v=mfQ3QVyw4N0
>
> And Robert Kosara's talk, *Presentation and Audience*, as part of his Advanced Visualization course for Observable, from 43:43 minutes in:
>
> - https://www.youtube.com/watch?v=Wb6xKQRtWig

## 7.3   Techniques

The technical element demonstrates how some of the uncertainty estimate examples in the chapter can be reproduced. We will again make use of functional programming approaches via the `purrr` package, mostly for generating and working over bootstrap resamples.

### 7.3.1   Import

- Download the `07-template.qmd`[1] file for this chapter, it to your `vis4sds` project.
- Open your `vis4sds` project in RStudio, and load the template file by clicking `File > Open File ... > 07-template.qmd`.

The template file lists the required packages: `tidyverse`, `sf`, `tidymodels` (for working with the bootstraps), `ggdist` and `distributional` for generating plots of parameter uncertainty and `gganimate` for the hypothetical outcome plot. Code for loading the STATS19 pedestrian crash data is in the `07-template.qmd` file.

### 7.3.2   Plot icon arrays

Icon arrays can be generated reasonably easily in standard ggplot2 using `geom_tile()` and some data generation functions. The most straightforward approach is to place icon arrays in a regularly-sized grid. In the example, KSI rates in Fareham (41%) and Oxford (17%) are compared.

First we generate the array data: a data frame of array locations (candidate crashes) with values representing whether the crash is slight or KSI depending on the observed KSI rate. In the code below, we set up a 10x10 grid of row and column locations and populate these with values for the selected local authorities (Oxford and Fareham) using base R's `sample()` function.

---

[1]https://vis4sds.github.io/vis4sds/files/07-template.qmd

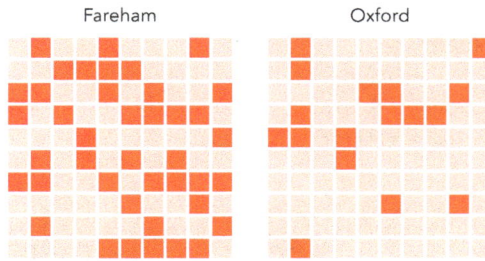

**Figure 7.11:** Icon arrays of pedestrian-vehicle crashes.

```
array_data <- tibble(
  row=rep(1:10, times=1, each=10),
  col=rep(1:10, times=10, each=1),
  Oxford=
    sample(
      c(rep(TRUE, times=1, each=17), rep(FALSE, times=1, each=83)),
      size=100, replace=FALSE),
  Fareham=
    sample(
      c(rep(TRUE, times=1, each=41), rep(FALSE, times=1, each=59)),
      size=100, replace=FALSE)
)
```

The plot code is straightforward:

```
array_data |>
  pivot_longer(
    cols=c(Oxford,Fareham), names_to="la", values_to="is_ksi"
    ) |>
  ggplot(aes(x=row,y=col, fill=is_ksi)) +
  geom_tile(colour="#ffffff", linewidth=1) +
  scale_fill_manual(values=c("#fee0d2","#de2d26"), guide="none") +
  facet_wrap(~la)
```

Plot specification:

1. *Data*: The array data, with `pivot_longer()` so that we can facet by local authority.
2. *Encoding*: x- and y-position according to the array locations and filled on whether the sampled crash is KSI or slight.
3. *Marks*: `geom_tile()` for drawing square icons.

4. *Scale*: `scale_fill_manual()` is supplied with values that are dark (KSI) and light (slight) red.
5. *Facets*: `facet_wrap()` for faceting on local authority.
6. *Setting*: Tiles are given large, white borders (`geom_tile(colour="#ffffff", size=1)`).

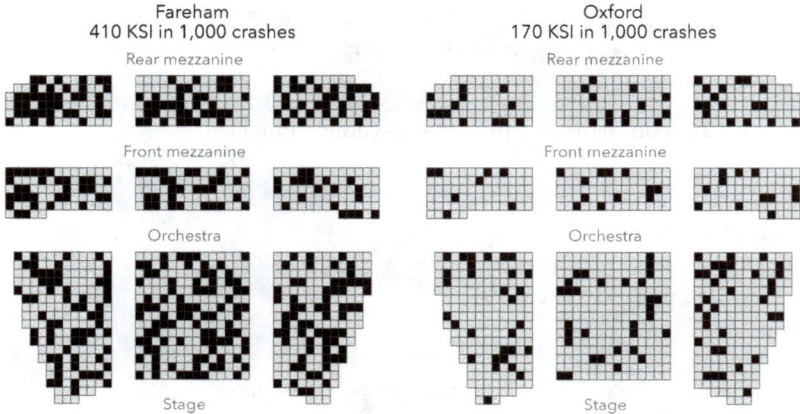

**Figure 7.12:** Risk theatre for pedestrian-vehicle crashes.

To present the icon array as a risk theatre, we have created a shapefile containing 1,000 theatre seat positions. To randomly allocate KSIs to seats on the proportion in which those crashes occur, we use the `slice_sample()` function.

```
theatre_cells <- st_read(here("data", "theatre_cells.geojson"))

ksi_seats <- bind_rows(
  theatre_cells |> slice_sample(n=170) |>
    add_column(la="Oxford\n170 KSI in 1,000 crashes"),
  theatre_cells |> slice_sample(n=410) |>
    add_column(la="Fareham\n410 KSI in 1,000 crashes")
)
```

The code:

```
theatre_cells |>
  ggplot() +
  geom_sf() +
  geom_sf(
    data=ksi_seats,
    fill="#000000"
```

```
) +
annotate("text", x=23, y=1, label="Stage", alpha=.5) +
annotate("text", x=23, y=21, label="Orchestra", alpha=.5) +
annotate("text", x=23, y=31, label="Front mezzanine", alpha=.5) +
annotate("text", x=23, y=42, label="Rear mezzanine", alpha=.5) +
facet_wrap(~la)
```

Plot specification:

1. *Data*: `theatre_cells` contains geometry data for all 1,000 seats; `ksi_seats` contains the randomly sampled seat locations.
2. *Marks*: `geom_tile()` for drawing seat icons.
3. *Facets*: `facet_wrap()` for faceting on local authority.
4. *Setting*: KSI tiles are coloured black (`fill="#000000"`). Also `annotate()` blocks of the theatre, `x`- and `y`- placement is determined via trial-and-error.

## 7.3.3 Generate bootstrap estimates of parameter uncertainty

The code for generating bootstrap resamples, stored in `rate_boots`, initially looks formidable. It is a template that is nevertheless quite generalisable, and so once learnt can be extended and applied to suit different use cases.

```
rate_boots <- ped_veh |>
  mutate(
    is_ksi=accident_severity!="Slight",
    year=lubridate::year(date)
  ) |>
  filter(year==2019,
         local_authority_district %in% c("Bristol, City of",
         "Sheffield", "Bromsgrove", "Cotswold")
  ) |>
  select(local_authority_district, is_ksi) |>
  nest(data=-local_authority_district) |>
  mutate(la_boot=map(data, bootstraps, times=1000, apparent=TRUE)) |>
  select(-data) |>
  unnest(la_boot) |>
  mutate(
    is_ksi=map(splits, ~analysis(.) |> pull(is_ksi)),
    ksi_rate=map_dbl(is_ksi, ~mean(.x)),
    sample_size=map_dbl(is_ksi, ~length(.x))
```

```
) |>
  select(-c(splits, is_ksi))
```

Code description:

1. *Setup*: The first `mutate()` is straightforward – a binary `is_ksi` variable identifies whether a crash is KSI, and the crash year is extracted from the `date` variable. Crashes recorded in 2019 are then filtered, along with the four comparator local authorities. To generate bootstrap resamples for each local authority, we `nest()` on local authority. You will remember that `nest()` creates a special type of column (a `list-column`) in which the values of the column is a list of data frames – in this case the crash data for each local authority. So running the code up to and including the `nest()`, a data frame is returned which contains four rows corresponding to the filtered local authorities and a `list-column` called `data`, each element of which is a data frame of varying dimensions (lengths) depending on the number of crashes recorded in each local authority.

2. *Generate bootstraps resamples*: In the `mutate()` that follows, `purrr`'s `map()` function is used to iterate over the list of datasets and the `bootstraps()` function to generate 1,000 bootstrap resamples for each nested dataset. The new column, `la_boot`, is a `list-column` this time containing a list of bootstrap datasets.

3. *Calculate sample estimates*: We `unnest()` the `la_boot` column to return a dataset with a row for each bootstrap resample and a `list-column` named `splits` which contains the bootstrap data. Again we `map()` over each element of `splits` to calculate the `ksi_rate` for each of the bootstrap datasets. The first call to `map()` extracts the `is_ksi` variable; the second is just a convenient way of calculating a rate from this (remembering that `is_ksi` is a binary variable); the third collects the sample size for each of the bootstraps, which of course is the number of crashes recorded for each local authority.

### 7.3.4   Plot parameter estimates with uncertainty information

With `ggdist`, the code for generating KSI rates with estimates of parameter uncertainty is straightforward and very similar to the error bar plots in the previous chapter.

Plot code:

```
rate_boots |>
  group_by(local_authority_district) |>
  mutate(std.error=sd(ksi_rate)) |>
  filter(id=="Apparent") |>
  ggplot(
    aes(x=reorder(local_authority_district, ksi_rate), y=ksi_rate)
    ) +
  stat_gradientinterval(
    aes(dist = dist_normal(mu=ksi_rate, sigma=std.error)),
    point_size = 1.5
  ) +
  coord_flip()
```

Plot specification:

1. *Data*: The `rate_boots` data frame is grouped by local authority and in the `mutate()` we calculate an estimate of bootstrap standard error, the standard deviation of the sampling distribution, and filter all rows where `id=="Apparent"` – this contains the KSI rate for the observed (unsampled) data.
2. *Encoding*: x- position varies according to local authority and y-position according to KSI rate. The estimated KSI rate and bootstrap standard error are also passed to `stat_gradientinterval()`, the `ggdist` function for producing gradient plots.
3. *Marks*: `stat_gradientinterval()` for drawing the gradients and point estimates.
4. *Setting*: `coord_flip()` for easy reading of local authority names.

## 7.3.5 Ensemble plots and hypothetical outcome plots

To generate bootstrap resamples on local authority and year, necessary for the year-on-year analysis, we can use the same template as that for calculating `rate_boots`; the only difference is that we `select()` and `nest()` on the `year` as well as the `local_authority_district` column.

```
rate_boots_temporal <- ped_veh |>
  ...
  ... |>
  select(local_authority_district, is_ksi, year) |>
  nest(-c(local_authority_district, year)) |>
  ...
```

```
...
...
```

The ensemble plot is again reasonably straightforward:

```
rate_boots_temporal |>
  ggplot(aes(x=year, y=ksi_rate)) +
  geom_line(data=. %>% filter(id=="Apparent"),
    aes(group=id), linewidth=.5) +
  geom_line(
    data=. %>% filter(id!="Apparent"),
    aes(group=id), alpha=.1, size=.2
    ) +
  facet_wrap(~local_authority_district)
```

Plot specification:

1. *Data*: The `rate_boots_temporal` data frame. Note that we include two line layers, one with the observed data (`data=. %>% filter(id=="Apparent")`) and one with the bootstrap data (`data=. %>% filter(id!="Apparent")`).
2. *Encoding*: x- position varies according to year, y-position according to KSI rate.
3. *Marks*: `geom_line()` for drawing lines.
4. *Facets*: `facet_wrap()` for faceting on local authority.
5. *Setting*: The bootstrap lines are de-emphasised by making the `alpha` and `size` channels very small.

Finally, the Hypothetical Outcome Plot (HOP) can be created easily using the `gganimate` package, simply by adding a call to `transition_states()` at the end of the plot specification:

```
rate_boots_temporal |>
  filter(id!="Apparent") |>
  ggplot(aes(x=year, y=ksi_rate)) +
  geom_line(aes(group=id), linewidth=.6) +
  facet_wrap(~local_authority_district)+
  transition_states(id, 0,1)
```

## 7.4 Conclusions

Uncertainty is fundamental to any data analysis. Statisticians and data scientists almost always end up reasoning about uncertainty, developing quantitative estimates of uncertainty and communicating uncertainty so that it can be taken into account when making evidence-based claims and decisions. Through an analysis of injury severity in the STATS19 road crash dataset, this chapter introduced techniques for quantifying and visually representing parameter uncertainty. There has been much activity in the Information Visualization and Data Journalism communities focussed on uncertainty communication – on developing approaches that promote intuition and allow users to experience uncertainty. We have covered some of these and demonstrated how they could be incorporated into our road crash analysis case study.

## 7.5 Further Reading

An excellent primer on uncertainty visualization:

* Padilla, L., Kay, M. and Hullman, J. 2021. "Uncertainty Visualization," in Wiley StatsRef: Statistics Reference Online, edited by B. Everitt N. Balakrishnan T. Colton and J. L. Teugels, *Wiley*. doi: 10.1002/9781118445112.stat08296.

On visualizing parameter uncertainty:

* Correll, M. and Gleicher, M. 2014. "Error Bars Considered Harmful: Exploring Alternate Encodings for Mean and Error," *IEEE Transactions on Visualization and Computer Graphics* 20(12): 2142–2151. doi: 10.1109/TVCG.2014.2346298.

* Kale, A., Nguyen, F., Kay, M. and Hullman, J. 2019. "Hypothetical Outcome Plots Help Untrained Observers Judge Trends in Ambiguous Data," *IEEE Transactions on Visualization and Computer Graphics*, 25(1): 892–902. doi: 10.1109/TVCG.2018.2864909.

On bootstrap resampling with R and tidyverse:

* Ismay, C. and Kim, A. 2020. "Statistical Inference via Data Science: A ModernDive into R and the Tidyverse", New York, NY: *CRC Press*. doi: 10.1201/9780367409913.
    – Chapters 7, 8.

# 8

## *Visual Storytelling*

By the end of this chapter you should gain the following knowledge and practical skills.

---

**i** Knowledge

☐ Appreciate the main characteristics of data-driven stories.
☐ Identify how visual and rhetorical devices are used to communicate with data.

---

**i** Practical skills

☐ Use shape primitives to code up custom chart designs in ggplot2.
☐ Add non-standard annotations to ggplot2 graphics.

---

## 8.1   Introduction

It is now taken for granted that we live in an evidence-based society in which data are deeply embedded in most domains. This recognition has coincided with the open source movement, which has freed up access and accelerated the development of tools for working with data. The response to Covid-19 is an excellent example. Enter *Covid19 github* into a search, and you'll be confronted with hundreds of code repositories demonstrating how data related to the pandemic can be collected, processed and analysed. This is exciting and feels very democratic. But there is a responsibility amongst those constructing and sharing evidence-based arguments to do so with integrity; navigating the difficult tension between communicating a clear message – necessarily reducing some of the complexity – at the same time as acknowledging uncertainty.

The role of narrative and storytelling when working with data is much discussed in Information Visualization (see Henry Riche et al. 2018) and Science Communication (see Franconeri et al. 2021). Importantly, this work recognises that there is no single, optimal visualization design that exposes the true structure or story in a dataset. Instead, careful design decisions must be made in light of data, audience and intended purpose. In this chapter we will review some of this literature with a special focus on approaches to communicating data around the Covid-19 pandemic, specifically publicly reported numbers of cases, hospitalisations and deaths.

## 8.2 Concepts

### 8.2.1 Data-driven storytelling

In earlier chapters of the book (e.g. Chapter 3) we identified and explored some common characteristics of effective data graphics. Roth (2021) enumerates 10 such characteristics specialised to data storytelling. Particularly important for visualization design is that data graphics are:

- *Designed*: The analyst makes very deliberate decisions in light of audience and purpose. The goal of visual storytelling is not just to *show* but also to *explain*.

- *Partial*: Essential information is prioritised and made salient, with abstraction and brevity preferred over complexity and completeness.

- *Intuitive*: Visual narratives take advantage of our natural tendency to communicate via metaphor and story, with a clear entry point and clear progression.

- *Compelling*: Visual stories often capture attention through an array of graphical devices – sequence, animation and interaction. They generate an aesthetic response.

- *Relatable and situated*: Visual stories promote empathy, using devices that place the audience in the story setting. They are usually constructed from somewhere – according to a particular position.

- *Political*: Visual data stories promote with clarity particular voices, interpretations or positions.

In the sections that follow, we review some prominent Covid-19 visualizations and reflect on how they implement these sorts of storytelling devices.

### 8.2.2 Designed and partial

Perhaps the most high-profile example of data graphics specialised to communication were those produced by the Financial Times (Financial Times 2020). Figure 8.1 displays one such example, the Financial Times' Covid-19 trajectory tracker. Along the y-axis are cumulative numbers of deaths represented using a log-scale; and along the x-axis, the number of days that elapsed since some threshold number of deaths was recorded.

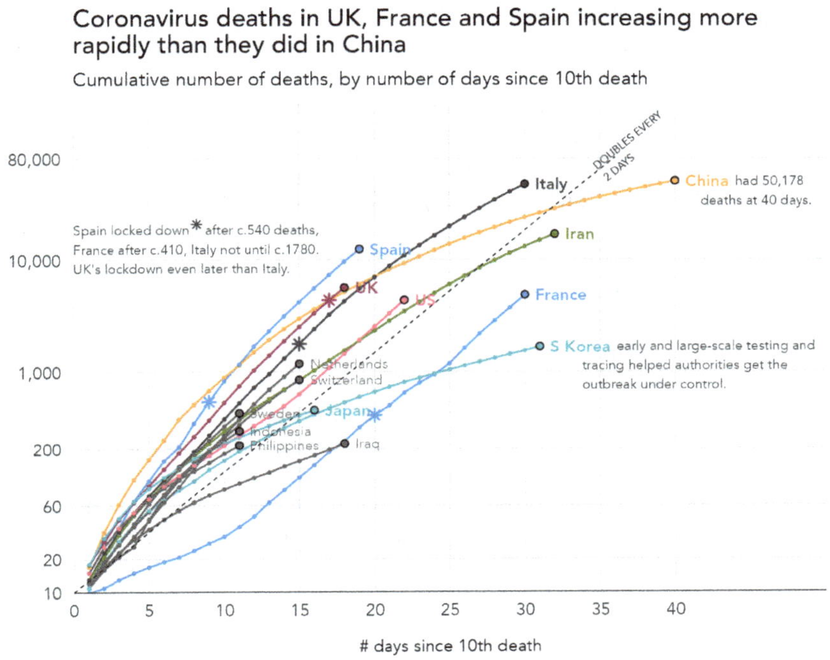

**Coronavirus deaths in UK, France and Spain increasing more rapidly than they did in China**

Cumulative number of deaths, by number of days since 10th death

**Figure 8.1:** John Burn-Murdoch's international comparison of deaths, as explained in Financial Times (2020). This is an approximate ggplot2 re-implementation of the original Financial Times graphic.

We can evaluate the graphic using some of the principles introduced in Chapter 3. In its use of position on an aligned scale to encode death counts, colour hue to differentiate countries and lines to connect daily counts, the graphic's visual encoding exploits our cognitive abilities. The graphic also makes appropriate use of superposition to support comparison (Gleicher et al. 2011) – lines for each country represented on the same coordinate space.

More interesting is how the graphic performs on Roth's (Roth 2021) characteristics of visual storytelling. It is clear that the graphic is *designed* with a very deliberate purpose:

1. Between country comparison: *Are countries on the same course?*
2. Comparison against milestone: *How many days does it take a certain county to reach a given number of deaths?*

It is possible to see each of these goals informing the graphic. Comparison between countries is most obviously supported by the use of a log-scale. This data transformation removes the dominant, hockey-stick type pattern inevitable when analysing disease growth (e.g. exponential doubling) and instead allows slopes, *growth rates*, to be compared directly. In narrowing on growth rates at the expense of absolute numbers, the graphic is *partial*. Even for readers of the Financial Times, though, log scales may not be so familiar. Annotations are therefore provided to anchor the reader on reference slope gradients (growth rates), again narrowing and supporting *intuition* on the essential goal of between country comparison. Notice also that there are no legends in this chart. Countries are differentiated with colour hue and then directly-labelled at their most recent death count, an addition that offloads an otherwise taxing look-up task, but also serves to emphasise a country's 'stage' in the pandemic.

> **i** Log scales for rate comparison
>
> A design alternative that supports between country comparison is Bhatia and Reich's (Bhatia and Reich 2020) Covid Trends chart (Figure 8.2). In this example, a double log scale is used, and growth rates in new cases are presented on the y-axis with total case numbers, rather than time elapsed, plotted along the x-axis. While the introduction of a double log scale might be judged to increase difficulty, actually this design narrows or simplifies the reader's visual judgement further on the thing that we are most interested in: comparison of country growth rates against the two day doubling (annotated with the diagonal). The chart is also accompanied with an excellent explanatory video, in which many of the characteristics of visual data stories enumerated by Roth (2021) can be identified.
>
>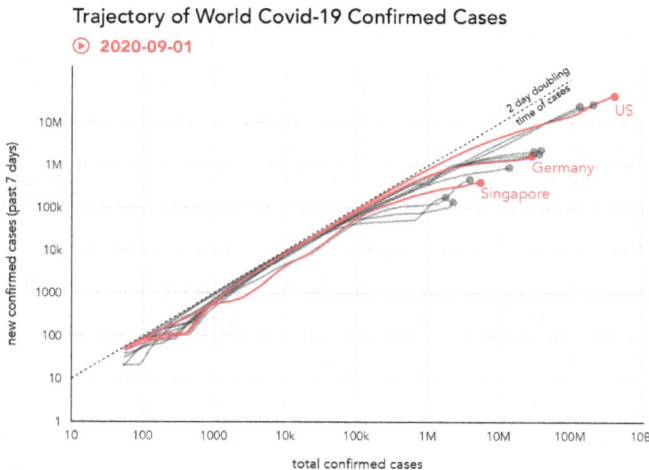
>
> **Figure 8.2:** Covid Trends chart (Bhatia and Reich 2020). This is an approximate ggplot2 re-implementation of a static from the original webpage.

### 8.2.3 Intuitive and compelling

Visual data stories are often explanatory (Roth 2021). They make *compelling* use of graphical and rhetorical devices to support understanding. This is especially important in data-driven storytelling, as often quite challenging concepts are covered in a limited amount of space. In Figure 8.3 is a static image from a data story written by Flourish (Lawlor and Robertson 2021) based on design work by Marteen Lambrechts. The data story is essentially a design exposition (Beecham, Dykes, Rooney, et al. 2021; Wood, Kachkaev, and Dykes 2018), guiding readers from the familiar to the unfamiliar. First a standard time series chart of hospitalisations and deaths is presented. Deficiencies in this layout are explained before progressively introducing the transformations involved to generate the preferred graphic, a connected scatterplot (Haroz, Kosara, and Franconeri 2016). Ordering the story in this way means that design decisions and trade-offs are covered from a familiar starting point, and from here new, sometimes unfamiliar encodings. Thinking about Roth's(2021) characteristics of visual storytelling, this approach to formulating a design story helps build *intuition*; there is a clear entry point and clear progression.

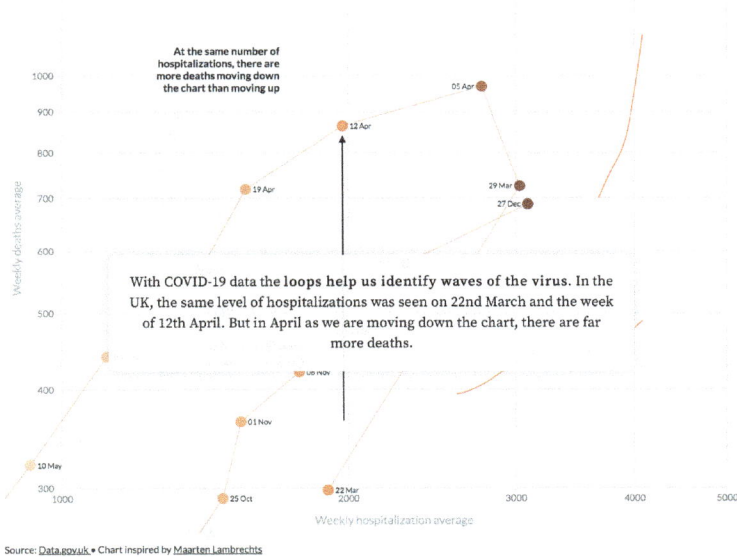

**Figure 8.3:** Screenshot from a data story written by Lawlor and Robertson (2021), demonstrating how connected scatterplots can be used to analyse changes in hospitalisations and deaths. The screenshot is reprinted, with permission, from *Flourish*.

> **i** Design exposition
>
> For an example of guided design exposition, see Beecham, Dykes, Hama,
> et al. (2021) which develops ways of showing simultaneously absolute
> and relative change in Covid-19 cases, with geographic context.

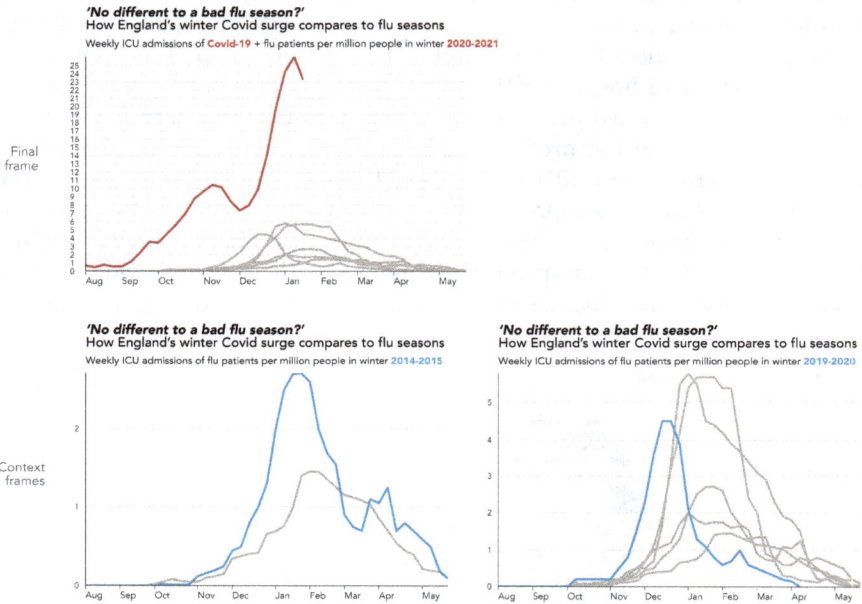

**Figure 8.4:** Weekly admissions to intensive care units in England. Each
year is introduced progressively, with animated rescaling of the y-axis used
to demonstrate how different in terms of intensive care admissions 2020/21
is to previous flu seasons. These are frames from an approximate ggplot2
re-implementation of the original Financial Times graphic (Financial Times
2021).

Animations in data graphics can increase engagement and aesthetic appeal.
They can also overwhelm since they involve complex tracking of information
between frames. An example of how animation can be used selectively to
build *intuition* is demonstrated in Figure 8.4, again from the Financial Times
(Financial Times 2021). The main objective is to demonstrate how different
2020-21 is in terms of admissions to intensive care compared to a normal year.
This was in response to claims that Covid-19 behaves much like seasonal flu; to
this extent the graphic is also quite *political*. Each year from 2013-14 is added
to the chart and the y-axis rescaled to reflect the new numbers. The animated

transitions of the y-axis help build expectation around normal variability in a similar way to the hypothetical outcome plots (Hullman, Resnick, and Adar 2015) covered in the previous chapter. The claim that the 2020-21 flu season is consistent with normal variability is then debunked by introducing the 2020-21 line in red, with animated rescaling of the y-axis used to further emphasise this point.

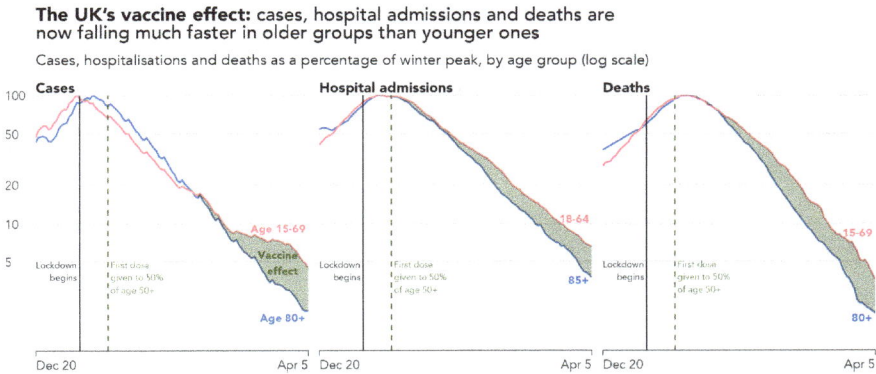

**Figure 8.5:** Data graphic by Burn-Murdoch (2021), presenting an argument for the role of vaccines in reducing Covid cases, hospital admissions and deaths. An approximate ggplot2 re-implementation of the original Financial Times graphic. Note that a slightly different smoothing function may have been used in the original graphic, especially for the cases data.

## 8.2.4   Political

Figure 8.5 presents a final example from the Financial Times (Burn-Murdoch 2021) with a clear *political* purpose. The graphic was created in response to some claims that it is movement restrictions (lockdowns) rather than vaccination that reduced infection rates in the country.

Interesting here is how annotation and *visual saliency* are used to direct our reading. If it was only annotated with points in time when lockdown and vaccination were initiated, the graphic might invite judgements about the effects of these two events on infectious rates. That it makes highly salient via annotation the (unmeasurable) *effect* of the vaccine is an interesting addition; there is little room for ambiguity.

This sort of presentation, labelling the chart with an unmeasurable vaccine effect, may risk graphical integrity. In a keynote given at IEEE VIS 2020, John Burn-Murdoch (2020), who created this and the other Financial Times graphics, reflected on his experiences generating data stories early in the pandemic. One observation was that the way in which the Financial Times's analyses were interpreted, and misinterpreted, varied depending on the prior expectations and political beliefs of those consuming them. The chart labelling in this case might have been added to signal more directly the evidence-based argument that was being made – to amplify a particular conclusion supported by the data.

Clearly all data analysis is heavily informed by the values, prejudices, motivations and incentives of those undertaking them. As demonstrated by Roth's (2021) characteristics of data storytelling, these individual factors are necessary to communicate effectively. An interesting question, then, is around how integrity and trust are designed into a data analysis. Perhaps in the case of the Financial Times data journalists, this trust is established over time, through a portfolio of data analysis work that is considered; communicated transparently and with humility.

---

**i** Subjectivity in design: revisiting travel-to-work under lockdown

When adhering strictly to graphical perception literature, there is a risk of assuming that *an* optimal design solution exists for any analysis situation – and that this optimal design can be achieved through correct mapping of data to visual encoding channels. The subtext to this chapter is that in reality we need to think about design in quite a subjective and context specific way.

Figure 8.6 uses the origin-destination travel-to-work data described in Chapter 4, updated with 2021 Census data, recorded at a time when a national Covid-19 lockdown and stay-at-home order was instituted. While the graphic is conventional in its visual encoding, it is constructed with a clear ambition: to highlight inequalities in the ability of workers to obey the stay-at-home order. This message could have been communicated easily with a single graphic, or even a single statistic, reporting proportional change between the two census years. However, to emphasise that professional high-status workers are better able to observe the 'stay-at-home' order than those in lower-status elementary occupations, the graphic is repeated many times for each London borough, with essentially the same trend. Figure 8.6 is therefore inefficient if judged against the detail it exposes. The repetition by London boroughs nevertheless has value in helping to reinforce a differences in experience, depending on occupation status, that appears to be universal across London. There is also a little creative license in the title: the data do not directly describe stay-at-home, since it is of course possible that workers commute to work inside the borough in which they live.

For the interested reader, the code can be found at:
`https://github.com/rogerbeecham/census21`.

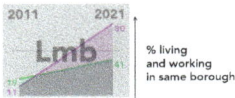

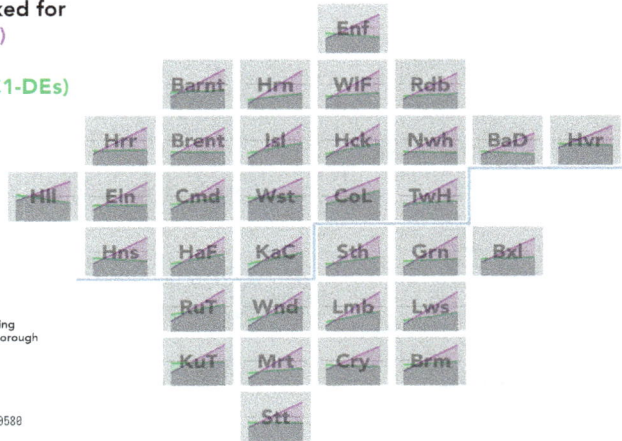

**Figure 8.6:** Gridmap showing proportions of those living and working in the same London borough in 2011 and 2021 Censuses.

---

**Task**

Watch John Burn-Murdoch's talk, *Making charts that make an impact*. John also presents a slightly different but persuasive take in this talk on uncertainty visualization:

- `https://www.youtube.com/watch?v=tIbaQUo6H9g`

And Robert Kosara's talk, *Presentation and Audience*, as part of his Advanced Visualization course for Observable, from `13:30` minutes in:

- `https://www.youtube.com/watch?v=Wb6xKQRtWig`

---

## 8.3 Techniques

The technical element demonstrates how to design plots deliberatively with annotations in ggplot2. We will recreate a glyphmap type graphic that originally appeared in The Washington Post (Thebault and Hauslohner 2020) to tell a story of growth in Covid-19 cases by US county. The graphic is presented in Figure 8.7. Each US county is represented as a line showing daily growth rates in new cases between 3rd May and 26th May 2020. Lines are positioned at the geographic centre of each county.

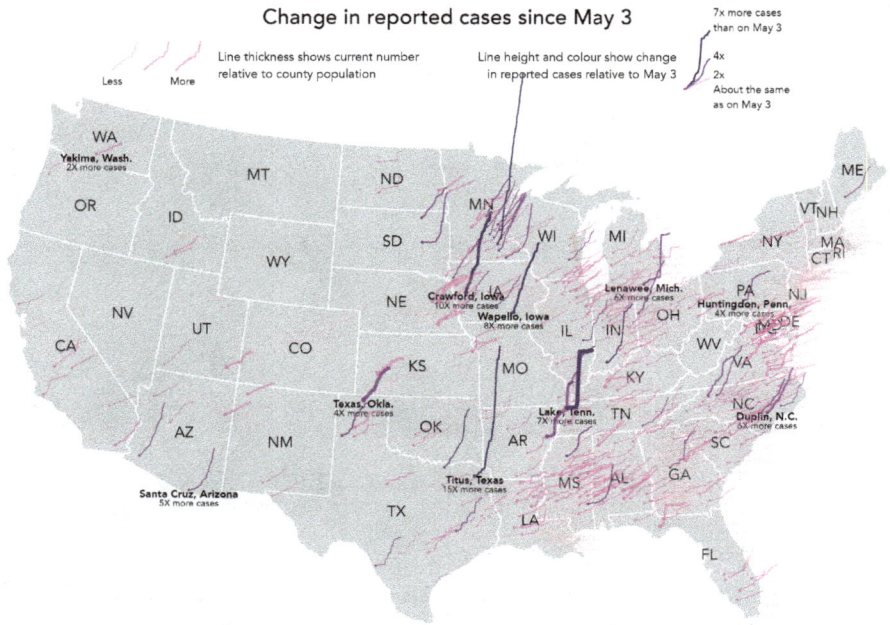

**Figure 8.7:** Glyphmap design displaying growth in COVID-19 cases by US county, based on the design by Thebault and Hauslohner (2020), originally in The Washington Post.

Figure 8.7 is certainly data dense. Without careful decisions on which aspects to emphasise it would be quite unreadable. Line thickness is varied according to relative infection rates (cumulative cases/population size), and growth rate is double encoded with colour value – darker and steeper lines for higher growth rates. Even with these additions it is challenging to discern trajectories for every county, but instead a typical model or expectation of these trajectories can be learnt from visually scanning the graphic. That there is spatial autocorrelation in trajectories means an overall pattern of exposure can be inferred, before eyes are drawn to exceptions. Initially these are towards the extreme end: tall, steep, dark and thick lines suggesting rapid growth rates and high case rates. Secondary patterns can also be observed, for example thick and mid-dark lines surrounded by lines that are generally lighter and thinner: counties that appear locally exceptional in having comparatively high growth and exposure rates.

The design is impressive, and there is an obvious benefit to showing growth rates in their spatial position. However, we are not looking at absolute numbers. The counties that are most salient are not those with the largest case counts. Rather, they have experienced rapid growth since the number of cases reported on 3rd May. So the graphic is most certainly *partial* and *designed* to suit a particular purpose. A slight adjustment in the implementation in Figure 8.7

was to only show growth rates for counties that had non-negligible case counts on 3rd May ($\geq$ 20 cases).

Without the careful integration of annotations and non-standard legends, Figure 8.7 would not be so successful. The aim of this technical section is to demonstrate an approach to generating heavily designed annotations – custom legends, which are often necessary when communicating with maps. For more extensive demonstration of how charts can be annotated and refined, do see the Further Reading section (Section 8.5) of this chapter.

### 8.3.1  Import

- Download the `08-template.qmd`[1] file, and save it to your `vis4sds` project.
- Open your `vis4sds` project in RStudio, and load the template file by clicking `File` > `Open File ...` > `08-template.qmd`.

The template file lists the required packages – `tidyverse` and `sf`. The data were collected using the `covdata` package (Healy 2020), attributing the county-level cumulative cases dataset maintained by data journalists at The New York Times (2021).

The template provides access to a version of this dataset that is 'staged' for charting. For this, cases are filtered on the dates covered by the Washington Post graphic (3rd to 25th May). Counties whose daily case counts were $\geq$ 20 on 3rd May are identified; calculated daily growth rates are anchored to case counts on 3rd May; 'end' growth rates and daily counts for each county are calculated (those recorded on 25th May); and finally a binned growth rate variable, identifying counties with daily case counts on 25th May that were $\leq 2\times$, $\geq 2\times$, $\geq 4\times$, $\geq 7\times$ those measured on 3rd May. Also there is a `state_boundaries` dataset to download, which contains `geometry` data for each US state, collected from US Census Bureau.

### 8.3.2  Plot trajectories

The main graphic is reasonably straightforward to construct. Different from many of the data graphics in earlier chapters, the way in which growth lines are coded is somewhat low-level. Remembering that lines are initially positioned in x- and y- on their county centroid, we generate from the data positions in geographic space for each observation – daily growth rates since 3rd May 2020. The code:

```
county_data |>
  ggplot() +
  geom_sf(
```

---

[1] https://vis4sds.github.io/vis4sds/files/08-template.qmd

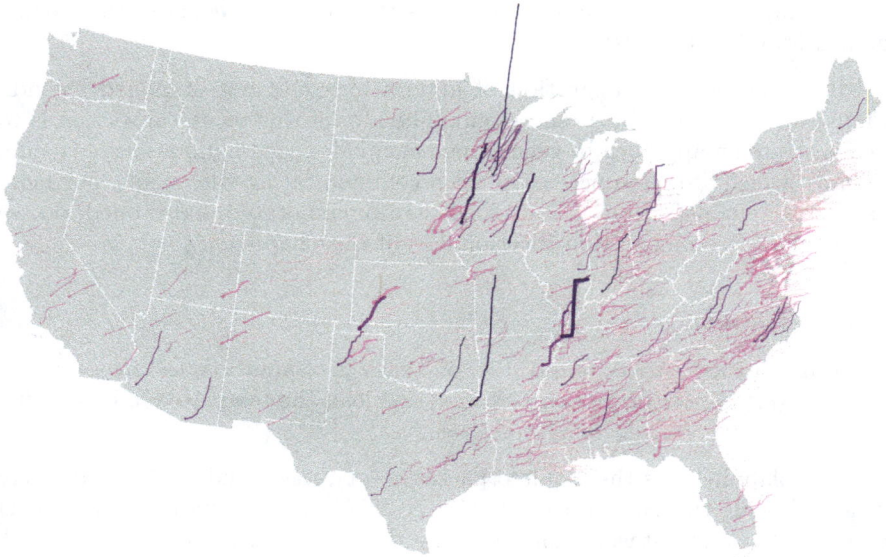

**Figure 8.8:** Glyphmap design displaying growth in COVID-19 cases by US county, without legend and annotations.

```
   data=state_boundaries,
   fill="#eeeeee", colour="#ffffff", linewidth=0.4
 )+
coord_sf(crs=5070, datum=NA, clip="off") +
geom_point(
  data=.%>% filter(date=="2020-05-03"),
  aes(
    x=x, y=y, size=case_rate, alpha=binned_growth_rate,
    colour=binned_growth_rate
    )
) +
# Plot case data.
geom_path(
  aes(
    x=x+((day_num-1)*6000), y=y+((growth_rate-1)*50000),
    group=fips, linewidth=case_rate, alpha=binned_growth_rate,
    colour=binned_growth_rate),
    lineend="round"
) +
scale_colour_manual(
  values=c("#fa9fb5", "#dd3497", "#7a0177", "#49006a")
```

```
  ) +
scale_size(range=c(.1,1.5)) +
scale_linewidth(range=c(.3,3)) +
scale_alpha_ordinal(range=c(.2,1)) +
guides(colour="none", size="none", alpha="none") +
theme_void()
```

The plot specification:

1. *Data*: The main dataset – the staged `county_data` file. Separately there is a `state_boundaries` file used to draw state boundaries and later label states. For the points drawn at the centroid of each US county (`geom_point()`), the data are filtered so that only a single day is represented (`filter(date=="2020-05-03")`).

2. *Encoding*: For `geom_point()`, x-position and y-position are mapped to county centroids (`x,y` variables in `county_data`), points are coloured according to `binned_growth_rate` using both `colour` and `alpha` and sized according to that county's `case_rate`. The same colour and size encoding is used for the lines (`geom_path()`). County lines are again anchored at county centroids but offset in `x` according to time elapsed (`day_num`) and in `y` according to `growth_rate`. The constants applied to `growth_rate` (5000) and `day_num` (6000), which control the space occupied by the lines, was arrived at manually through trial and error. Note that these numbers are large, as they relate to geographic coordinate space. In order to draw separate lines for each county, we set the `group=` argument to the county identifier variable `fips`.

3. *Marks*: `geom_point()` for the start points centred on county centroids and `geom_path()` for the lines.

4. *Scale*: `scale_colour_manual()` for the binned growth rate colours; `scale_alpha()` for an ordinal transparency range. The floor for this is 0.2 and not 0, otherwise counties with the smallest binned growth rates would not be visible. `scale_size()` and `scale_linewidth_size()` for varying the size of points and thickness of lines continuously according to case rate, the range was arrived at through trial and error.

5. *Setting*: We don't want the default legend to appear, and so `guides()` turns these off. Additionally `theme_void()` removes the default axes, gridlines etc.

### 8.3.3 Add labels and annotations

The two-letter state boundaries held in the `state_boundaries` file can be added in a `geom_text()` layer, positioned in `x` and `y` at state centroids. For obvious reasons this needs to appear *after* the first call to `geom_sf()`, which draws the filled state outlines:

```
county_data |>
  ggplot()+
  ggplot() +
  geom_sf(
    data=state_boundaries,
    fill="#eeeeee", colour="#ffffff", linewidth=0.4
  )+
  geom_text(data=state_boundaries, aes(x=x,y=y,label=STUSPS), alpha=.8)+
  ...
  ...
  ...
```

For the counties annotated with accompanying growth rates, we create a staged, filtered data frame containing only those counties and with just one row for each county. This is a little more tedious as we have to manually identify these in a `filter()`. Note that we filter on `date` first, so that only one row is returned for each county. Within the `mutate()` some manual abbreviations are made for state names and also the `end_rate` variable is rounded to whole numbers for better labelling.

```
# Counties to annotate.
annotate <- county_data |>
  filter(
    date=="2020-05-03",
    county==c("Huntingdon") & state=="Pennsylvania" |
    county==c("Lenawee") & state=="Michigan" |
    county==c("Crawford") & state=="Iowa" |
    county==c("Wapello") & state=="Iowa" |
    county==c("Lake") & state=="Tennessee" |
    county=="Texas" & state == c("Oklahoma") |
    county==c("Duplin") & state=="North Carolina" |
    county==c("Santa Cruz") & state=="Arizona"|
    county==c("Titus") & state=="Texas"|
    county==c("Yakima") & state=="Washington"
  ) |>
  mutate(
    state_abbr=case_when(
      state=="Pennsylvania" ~ "Penn.",
      state=="Iowa" ~ "Iowa",
      state=="Tennessee" ~ "Tenn.",
      state=="Oklahoma" ~ "Okla.",
      state=="Texas" ~ "Texas",
      state=="North Carolina" ~ "N.C.",
      state=="Washington" ~ "Wash.",
```

```
      state=="Michigan" ~ "Mich.",
      state=="Arizona" ~ "Arizona",
      TRUE ~ ""),
   end_rate_round = round(end_rate,0)
 )
```

Plotting these is again quite straightforward with `geom_text()`. The `paste0()` function is used to build labels that display county names (`county`) and then state abbreviations (`state_abbr`). These appear below each county by offseting y-position. Additionally the counties are given a bold font by passing an argument to `fontface="bold"`. The same approach is used for the rate labels, but with an incremented y-position offset so that they don't overlap the county name labels.

```
county_data |>
  ggplot()+
  geom_sf(data=state_boundaries,
    fill="#eeeeee", colour="#ffffff", linewidth=0.4)+
  ...
  ...
  geom_text(
    data=annotate,
    aes(x=x,y=y-20000,label=paste0(county,", ",state_abbr)),
    size=3, fontface="bold"
    )+
  geom_text(
    data=annotate,
    aes(x=x,y=y-65000,label=paste0(end_rate_round,"X more cases")),
    size=2.5
    )+
  ...
  ...
  ...
```

### 8.3.4 Build custom legend

Since Figure 8.7 is a custom data graphic coded in a low-level way with `geom_segment()`, it is useful to accompany it with a more expressive legend. We therefore build our own legend from scratch, using the geographic space of the plot as our canvas.

To support positioning of the legend we extract the spatial limits, or bounding box, of our plot area – the mainland US.

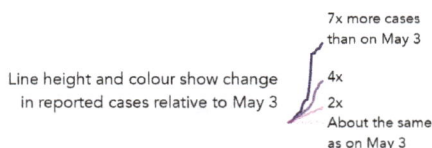

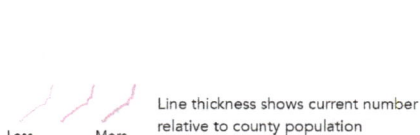

**Figure 8.9:** Growth rates legend.    **Figure 8.10:** Case rates legend.

```
# Bounding box for mainland US.
bbox <- st_bbox(state_boundaries)
width <- bbox$xmax-bbox$xmin
height <- bbox$ymax-bbox$ymin
```

We then create a dataset for the top right legend displaying the different categories of growth rate – Figure 8.9. Counties filtered by their different growth rates were identified manually. As you will see shortly, we use exactly the same *encoding* as the main graphic for the example legend lines, but rather than positioning these selected counties in their real geographic position, we override their x- and y- location so that the lines appear in a margin to the top right of the graphic. This is achieved in the `mutate()`, where we set x-position to start at the right quarter of the graphic (`bbox$xmax-.25*width`) and y-position to start slightly above the top of the graphic (`bbox$ymax+.05*height`). `case_rate` is set to a constant as we don't want line width to vary.

```
# Legend : growth
legend_growth <- county_data |>
  filter(
    county=="Dubois" & state=="Indiana" |
    county=="Androscoggin" & state=="Maine" |
    county=="Fairfax" & state=="Virginia" |
    county=="Bledsoe" & state=="Tennessee"
  ) |>
  mutate(
    x=bbox$xmax-.25*width, y=bbox$ymax+.05*height,
    case_rate=.01,
    label=case_when(
      county == "Dubois" ~ "7x more cases than on May 3",
      county == "Androscoggin" ~ "4x",
      county == "Fairfax" ~ "2x",
      county == "Bledsoe" ~ "About the same as on May 3"
    )
  )
```

A separate dataset is also created for drawing the top left legend – Figure 8.10, showing different case rates relative to population size. In the `mutate()` we set x-position to start towards the left of the graphic (`bbox$xmax-.88*width`) and y-position to start slightly above the top of the graphic `bbox$ymax+.05*height`. We want to draw three lines corresponding to a low, medium and high growth rate and so `pivot_longer()` to duplicate the daily case data over rows. Each line is positioned with the `offset_day` variable, a multiple applied to the geographic `width` of US used later in the ggplot2 specification.

```
# Legend : case
legend_case <- county_data |>
  filter(county == "Kings" & state=="California" ) |>
  mutate(
    x=bbox$xmax-.88*width,y=bbox$ymax+.05*height,
    binned_growth_rate=factor(binned_growth_rate)
  ) %>%
  select(x, y, day_num, growth_rate, binned_growth_rate, fips) |>
  mutate(low=.001, mid=.009, high=.015) |>
  pivot_longer(
    cols=c(low, mid, high), names_to="offset", values_to="offset_rate"
  ) |>
  mutate(
    offset_day= case_when(
      offset == "low" ~ 0,
      offset == "mid" ~ .04,
      offset == "high" ~ .08
    )
  )
```

## 8.3.5   Compose graphic

The code block below demonstrates how derived data for the legends are used in the ggplot2 specification. Exactly the same mappings are used in the legend as the main graphic, and so the call to `geom_path()` looks similar, except for the different use of x- and y- position. Labels for the legends are generated using `annotate()` and again positioned using location information contained in `bbox`.

```
# Text for annotations and titles.
growth_text <- "Line height and colour show change in reported cases
relative to May 3"
case_text <- "Line thickness shows current number relative to
county population"
title_text <- "Change in reported cases since May 3"
```

```
county_data |>
  ggplot()+
  geom_sf(data=state_boundaries,
    fill="#eeeeee", colour="#bcbcbc", linewidth=0.4)+
  ...
  ...
  ...
  # Plot growth legend lines.
  geom_path(
    data=legend_growth,
    aes(x=x+((day_num-1)*6000), y=y+((growth_rate-1)*50000),
      group=fips, linewidth=case_rate, alpha=binned_growth_rate,
      colour=binned_growth_rate),lineend="round") +
  # Text label for growth legend lines.
  geom_text(
    # For positioning manually edit growth_rate of Bledsoe.
    data=legend_growth %>%
      filter(day_num == max(county_data$day_num)) %>%
      mutate(growth_rate=if_else(county=="Bledsoe", -1,growth_rate)),
    aes(x=x+(day_num*6000)+10000,y=y+((growth_rate-1)*50000),
      label=str_wrap(label, 15)),
      size=2.5
      )+
  annotate("text",
    x=bbox$xmax-.25*width, y=bbox$ymax+.08*height,
    label=str_wrap(growth_text,35), size=3.5, hjust=1
    )+
  # Plot case legend lines.
  geom_path(
    data=legend_case, aes(x=x+((day_num-1)*6000)+offset_day*width,
      y=y+((growth_rate-1)*50000),
      group=paste0(fips,offset), linewidth=offset_rate,
      alpha=binned_growth_rate, colour=binned_growth_rate),
      lineend="round"
    ) +
  # Text label for case legend lines.
  annotate("text", x=bbox$xmax-.88*width, y=bbox$ymax+.04*height,
    label="Less", size=2.5)+
  annotate("text", x=bbox$xmax-.8*width, y=bbox$ymax+.04*height,
    label="More", size=2.5)+
  annotate("text",
    x=bbox$xmax-.75*width, y=bbox$ymax+.08*height,
    label=str_wrap(case_text,35), size=3.5, hjust=0
```

```
 )+
# Title.
annotate("text",
  x=bbox$xmax-.5*width, y=bbox$ymax+.15*height,
  label=title_text, size=5
 )+
...
...
...
```

## 8.4 Conclusions

Communicating effectively with data is not an easy undertaking. Difficult decisions must be made around how much detail to sacrifice in favour of clarity and simplicity of message. Visual approaches can help here, giving cues that order and prioritise information and that build explanatory narratives using metaphor and other rhetorical devices. There are stellar examples of this from in-house data journalism teams. We have considered some of these and the careful design decisions made when communicating data-driven stories in light of data, audience and intended purpose. Many data journalists use ggplot2 as their visualization toolkit of choice, and in the technical section we demonstrated how more *designed* graphics can be generated. This somewhat fiddly approach to creating graphics is different from the style of workflow envisaged in Chapter 3 and Chapter 4, on exploratory visual analysis. However, as demonstrated through the examples in this chapter and the book more generally, ggplot2 can be used for this more deliberative visualization design, making control over annotations, text labels and embedded graphics useful skills to develop.

## 8.5 Further Reading

Cédric Scherer's excellent workshop at posit::conf(2023). Highly recommended as a resource for covering in a methodical way how to parameterise ggplot2 scales, coordinate systems, facets, annotation and labelling.

- Cédric Scherer, 2023. "Designing Data Visualizations to Successfully Tell a Story"[2], A posit::conf(2023) Workshop.

---

[2]https://posit-conf-2023.github.io/dataviz-storytelling/

For a similarly excellent resource see:

- Kieran Healy, 2019. "Data Visualization: A Practical Introduction", Princeton, NJ: *Princeton University Press.*
  - Chapter 8

Data journalists at the BBC have assembled a useful 'cookbook' demonstrating basic but useful edits to standard ggplot2 graphics.

- BBC, 2019. "BBC Visual and Data Journalism cookbook for R graphics"[3].

High-level but principled description of how to incorporate annotations, customised scales and legends when generating plots for communication.

- Wickham, H., Çetinkaya-Rundel, M., Grolemund, G. 2023, "R for Data Science, 2nd Edition", Sebastopol, CA: *O'Reilly.*
  - Chapter 11.

---

[3]https://bbc.github.io/rcookbook/

# A

---

## Task Answers

---

> **i** Task answers
>
> This page contains answers to the *Task* activities presented in selected chapters of the book.

---

## A.1   From Chapter 2

### Task 1

The completed data description table for the Citibike stations dataset.

| Variable name | Variable value | Measurement level |
|---|---|---|
| name | "Central Park" | Categorical-nominal |
| capacity | 80 | Ratio |
| rank_capacity | 45 | Ordinal |
| date_opened | "2014-05-23" | Interval |
| longitude | -74.00149746 | Interval |
| latitude | 40.74177603 | Interval |

### Task 2

```
# # A tibble: 7 × 3
#    day    Customer Subscriber
#    <ord>     <dbl>      <dbl>
# 1 Sun       0.198      0.144
# 2 Mon       0.137      0.163
# 3 Tue       0.144      0.172
# 4 Wed       0.104      0.125
# 5 Thu      0.0973      0.122
```

```
# 6 Fri       0.135        0.138
# 7 Sat       0.185        0.136
```

Code for exploring the proportion of trips made by day of week on `user_type`:

```
ny_temporal |>
  group_by(day, user_type) |>
  summarise(count=sum(count)) |> ungroup() |>
  pivot_wider(names_from=user_type, values_from=count) |>
  mutate(across(.cols=c(Customer, Subscriber), .fns=~.x/sum(.x)))
```

## A.2   From Chapter 3

### Task 1

The completed encoding description table for the Washington Post election map (Figure 3.5).

| Data item | Measurement level | Visual mark | Visual channel | Rank |
|-----------|-------------------|-------------|----------------|------|
| County location | Interval | Lines | Position in x-y- | 1 mag:order |
| County winner | Cat-nominal | Lines | Colour hue | 2 id:category |
| County flip from 2012 | Cat-ordinal | Lines | 1D size | 3 mag:order |
| County Swing fom 2012 | Ratio | Lines | Tilt/angle | 4 mag:order |
| State 'winner' | Cat-nominal | Polygon | Colour hue | 2 id:category |

### Task 2

Code to produce a set of histograms similar to those in Figure 3.8.

```
data_gb |>
  ggplot(mapping=aes(swing_con_lab)) +
  geom_histogram(fill="#003c8f") +
  # Annotate with median Swing.
```

```
geom_vline(xintercept=4.44, size=.3)+
labs(x="Swing", y="count")+
facet_wrap(~region)
```

## Task 3

The code to reproduce the graphic in Figure 3.11.

```
con <- "#0575c9"
lab <- "#ed1e0e"
other <- "#bdbdbd"

data_gb |>
  mutate(is_flipped=seat_change_1719=="Conservative gain from Labour",
         is_flipped=if_else(is.na(is_flipped), FALSE, is_flipped),
         winner_19=case_when(
           winner_19 == "Conservative" ~ "Conservative",
           winner_19 == "Labour" ~ "Labour",
           TRUE ~ "Other"
         )) |>
  ggplot(aes(x=con_17, y=con_19)) +
  geom_point(aes(colour=winner_19, alpha=is_flipped,
    shape=is_flipped)) +
  geom_abline(intercept = 0, slope = 1, size=.3) +
  scale_colour_manual(values=c(con,lab,other)) +
  scale_alpha_ordinal(range=c(.5,1)) +
  scale_shape_manual(values=c(21,19)) +
  scale_x_continuous(limits=c(0,90)) +
  labs(x="vote share 2017 ", y="vote share 2019")
```

## A.3   From Chapter 4

## Task 1

The code to reproduce the heatmap in the left column of Figure 4.6.

```
# Vector of vehicles in order they appear in graphic.
# Convert vehicle_type to factor to effect ordering.
order_type <- c("Car", "Taxi", "Bus", "Motorcycle", "Other","Van",
  "HGV", "Bicycle")
```

```r
# For new is_inner variable used to facet plot.
inner_boroughs <- c("Camden", "Greenwich", "Hackney",
  "Hammersmith and Fulham", "Islington","Kensington and Chelsea",
  "Lambeth", "Lewisham", "Southwark", "Tower Hamlets", "Wandsworth",
  "Westminster", "City of London")
# Staged dataset where signed-chi residuals are created.
model_data <- ped_veh |>
  filter(police_force == "Metropolitan Police" |
    police_force == "City of London") |>
  mutate(
    is_inner=if_else(local_authority_district %in%
      inner_boroughs, "inner", "outer"),
    vehicle_type=factor(vehicle_type, levels=order_type)
  ) |>
  group_by(local_authority_district) |>
  mutate(row_total=n()) |> ungroup() |>
  group_by(vehicle_type) |>
  mutate(col_total=n()) |> ungroup() |>
  mutate(grand_total=n()) |>
  group_by(local_authority_district, vehicle_type) |>
  summarise(
    observed=n(), row_total=first(row_total),
    col_total=first(col_total), grand_total=first(grand_total),
    expected=(row_total*col_total)/grand_total,
    resid=(observed-expected)/sqrt(expected),
    is_inner=first(is_inner)
  ) |>  ungroup()
# Find max residual value to ensure colour scheme is symmetrical on 0.
max_resid <- max(abs(model_data$resid))

# Plot heatmap.
model_data |>
  ggplot(aes(x=vehicle_type,
    y=reorder(local_authority_district, row_total))) +
  geom_tile(aes(fill=resid), colour="#ffffff", size=.4) +
  facet_grid(is_inner~., scales="free_y", space="free_y") +
  scale_fill_distiller(palette="RdBu", direction=-1,
    limits=c(-max_resid,max_resid)) +
  guides(fill="none")
```

## Task 2

The code to reproduce the design challenge plots in Figure 4.12.

```r
# Calculate crash freqs by IMD class of location,
# dark/daylight and casulty age.
plot_data <- ped_veh |>
  filter(
    age_of_casualty>0, crash_quintile != "Data missing or out of range",
    light_conditions != "Data missing or out of range"
    ) |>
  mutate(is_daylight=factor(
    if_else(light_conditions == "Daylight", "daylight", "dark"),
              levels=c("dark", "daylight"))) |>
  group_by(age_of_casualty, is_daylight, crash_quintile) |>
  summarise(count=n()) |> ungroup()


# Top plot.
plot_data |>
  ggplot(aes(x=age_of_casualty, y=count)) +
  geom_col(aes(colour=is_daylight), width=1) +
  facet_grid(is_daylight~crash_quintile, space="free_y",
    scales="free_y", labeller=labeller(c("daylight", "dark"))) +
  scale_colour_manual(values=c("#08519c", "#c6dbef"), guide="none")+
  labs(y="crash count in hundreds", x="casualty age") +
  scale_y_continuous(
    breaks=c(c(2,4,6,8,10)*100),
    labels = scales::comma_format(scale = .01))


# Bottom plot.
plot_data |>
  # Calcuate % crashes in daylight and expected daylight counts
  # in each data item from this.
  mutate(total=sum(count)) |>
  pivot_wider(names_from=is_daylight, values_from=count) |>
  mutate(
    prop_daylight=sum(daylight, na.rm=TRUE)/first(total),
    expected_daylight=(daylight+dark)*prop_daylight
  ) |>
  pivot_longer(cols=c(dark, daylight),
    names_to="is_daylight", values_to="count") |>
  # Plot.
  ggplot(aes(x=age_of_casualty, y=count)) +
  geom_col(aes(colour=is_daylight), width=1) +
  geom_line(aes(y=expected_daylight, group=crash_quintile),
    colour="#737373", linewidth=.4) +
  facet_wrap(~crash_quintile,
    labeller=labeller(c("daylight", "dark")), nrow=1) +
```

```
scale_colour_manual(values=c("#08519c", "#c6dbef"), guide="none")+
labs(y="crash count in hundreds", x="casualty age") +
scale_y_continuous(
  breaks=c(c(4,8,12)*100),
  labels = scales::comma_format(scale = .01))
```

## A.4   From Chapter 5

### Task 1

- For jobs filled in the City of London (CoL), from which borough does the largest number of workers commute?
  - Answer: Wandsworth (Wnd)
- For jobs filled in Camden (Cmd), from which borough does the largest number of workers commute?
  - Answer: Barnet (Barnt), maybe Islington (Isl)
- Eyeballing the graphic, identify the top 3 boroughs which appear to have the most localised labour markets in terms of in-commuting.
  - Answer: Bexleyheath (Bxl), Havering (Hvr), Barking and Dagenham (BaD). Also, Sutton (Sttn), Grenwich (Grn) look very localised.

## A.5   From Chapter 6

The code to reproduce the annotated parallel-coordinate-plot in Figure 6.3.

```
# List of variables ordered ascending by correlation with leave.
order_vars <- cons_data |>
  mutate(across(c(younger:heavy_industry), ~(.x-mean(.x))/sd(.x))) |>
  pivot_longer(cols=younger:heavy_industry,
    names_to="expl_var", values_to="prop") |>
  group_by(expl_var) |>
  summarise(cor=cor(leave,prop)) |> ungroup() |> arrange(cor) |>
  pull(expl_var)

# Staged dataset of z-score transformed variables and variables
# identifying extreme Leave/Remain constituencies for highlighting.
plot_data <- cons_data |>
```

```
  mutate(
    majority=if_else(leave>.5, "Leave", "Remain"),
    across(c(leave, younger:heavy_industry), ~(.x-mean(.x))/sd(.x)),
    decile=ntile(leave, 10), is_extreme=decile > 9 | decile < 2
  ) |>
  select(
    majority, is_extreme, decile, constituency_name, leave, degree,
    professional,  younger, eu_born, no_car, white, own_home, christian,
    not_good_health, heavy_industry) |>
  # Change polarity in selected variables.
  mutate(degree=-degree, professional=-professional, younger=-younger,
    eu_born=-eu_born, no_car=-no_car) |>
  pivot_longer(cols= c(leave:not_good_health),
    names_to="var", values_to="z_score") |>
  # Explanatory variable as factor ordered according to known assocs
  # with Leave.
  mutate(
    var=factor(var, levels=c("leave", order_vars)),
    var=fct_rev(var)
  )
# Sample extreme constituencies, for Leave and Remain,
# each time plot is built.
annotate_data <- plot_data |>
  filter(is_extreme) |>
  group_by(decile) |>
  sample_n(1) |> pull(constituency_name)

plot_data |>
  ggplot(aes(x=var, y=z_score, colour=majority,
    group=c(constituency_name))) +
  geom_path(alpha=0.15, linewidth=.2) +
  # Highlight extreme remain/leave constituencies.
  geom_path(data= . %>%
    filter(constituency_name %in% annotate_data),
    alpha=1, linewidth=.4) +
  geom_text(
    data= . %>%
      filter(constituency_name %in% annotate_data, var=="leave"),
    aes(x="leave", y=z_score, label=str_wrap(constituency_name,15)),
    size=3.5, vjust="top", hjust="centre", nudge_x=+.5
    ) +
  scale_colour_manual(values=c("#b2182b","#2166ac")) +
  coord_flip()
```

# *References*

BBC Visual and Data Journalism Team. 2019. "BBC Visual and Data Journalism Cookbook for R Graphics." https://github.com/bbc/rcookbook.

Beecham, R. 2020. "Using Position, Angle and Thickness to Expose the Shifting Geographies of the 2019 UK General Election." *Environment and Planning A: Economy and Space* 52 (5): 833–36. https://doi.org/10.1177/0308518X 20909392.

———. 2024. "`gridmappr`: An R Package for Creating Small Multiple Gridmap Layouts." In *GISRUK 2024*. Leeds, UK: Zenodo. https://doi.org/10.5281/zenodo.10926863.

Beecham, R., J. Dykes, L. Hama, and N. Lomax. 2021. "On the Use of 'Glyphmaps' for Analysing the Scale and Temporal Spread of COVID-19 Reported Cases." *ISPRS International Journal of Geo-Information* 10 (4). https://doi.org/10.3390/ijgi10040213.

Beecham, R., J. Dykes, W. Meulemans, A. Slingsby, C. Turkay, and J. Wood. 2017. "Map Line-Ups: Effects of Spatial Structure on Graphical Inference." *IEEE Transactions on Visualization & Computer Graphics* 23 (1): 391–400. https://doi.org/10.1109/TVCG.2016.2598862.

Beecham, R., J. Dykes, C. Rooney, and W. Wong. 2021. "Design Exposition Discussion Documents for Rich Design Discourse in Applied Visualization." *IEEE Transactions on Visualization and Computer Graphics* 27 (8): 3451–62. https://doi.org/10.1109/TVCG.2020.2979433.

Beecham, R., and R. Lovelace. 2023. "A Framework for Inserting Visually-Supported Inferences into Geographical Analysis Workflow: Application to Road Safety Research" 55: 345–66. https://doi.org/10.1111/gean.12338.

Beecham, R., and A. Slingsby. 2019. "Characterising Labour Market Self-Containment in London with Geographically Arranged Small Multiples." *Environment and Planning A: Economy and Space* 51 (6): 1217–24. https://doi.org/10.1177/0308518x19850580.

Beecham, R., A. Slingsby, and C. Brunsdon. 2018. "Locally-Varying Explanations Behind the United Kingdom's Vote to Leave the European Union." *Journal of Spatial Information Science* 16: 117–36. https://doi.org/10.5311/josis.2018.16.377.

Beecham, R., N. Williams, and L. Comber. 2020. "Regionally-Structured Explanations Behind Area-Level Populism: An Update to Recent Ecological Analyses." *PLOS One* 15 (3): e0229974. https://doi.org/10.1371/journal.pone.0229974.

Beecham, R., and J. Wood. 2014. "Exploring Gendered Cycling Behaviours Within a Large-Scale Behavioural Data-Set." *Transportation Planning and Technology* 37 (1): 83–97. https://doi.org/10.1080/03081060.2013.844903.

Beecham, R., Y. Yang, C. Tait, and R. Lovelace. 2023. "Connected Bikeability in London: Which Localities Are Better Connected by Bike and Does This Matter?" *Environment and Planning B: Urban Analytics and City Science* 50 (8): 2103–17. https://doi.org/10.1177/23998083231165122.

Bhatia, A., and H. Reich. 2020. *Covidtrends*. https://github.com/aatishb/covidtrends.

Boukhelifa, N., A. Bezerianos, T. Isenberg, and J. Fekete. 2012. "Evaluating Sketchiness as a Visual Variable for the Depiction of Qualitative Uncertainty." *IEEE Transactions on Visualization and Computer Graphics* 18 (12): 2769–78. https://doi.org/10.1109/TVCG.2012.220.

Brewer, C., and A. Campbell. 1998. "Beyond Graduated Circles: Varied Point Symbols for Representing Quantitative Data on Maps." *Cartographic Perspectives*, no. 29: 6–25. https://doi.org/10.14714/CP29.672.

Brunsdon, C., and M. Charlton. 2011. "An Assessment of the Effectiveness of Multiple Hypothesis Testing for Geographical Anomaly Detection." *Environment and Planning B: Planning and Design* 38 (2): 216–30. https://doi.org/10.1068/b36093.

Brunsdon, C., and A. Comber. 2021. "Opening Practice: Supporting Reproducibility and Critical Spatial Data Science." *Journal of Geographical Systems* 23: 477–96. https://doi.org/10.1007/s10109-020-00334-2.

Brunsdon, C., M. Fortheringham, and M. Charlton. 2002. "Geographically Weighted Summary Statistics: A Framework for Localised Exploratory Data Analysis." *Computers, Environment and Urban Systems* 26: 501–24. https://doi.org/10.1016/S0198-9715(01)00009-6.

Buja, A., D. Cook, H. Hofmann, M. Lawrence, E-K Lee, D. Swayne, and H. Wickham. 2009. "Statistical Inference for Exploratory Data Analysis and Model Diagnostics." *Philosophical Transactions of the Royal Society A: Mathematical, Physical and Engineering Sciences* 367 (1906): 4361–83. https://doi.org/10.1098/rsta.2009.0120.

Burn-Murdoch, J. 2020. "BELIV 2020 Keynote: John Burn-Murdoch." In *2020 IEEE Workshop on Evaluation and Beyond - Methodological Approaches to Visualization (BELIV)*, 9–9. https://doi.org/10.1109/BELIV51497.2020.00007.

———. 2021. "Vaccines Are Working: Charts That Show the Covid Endgame." *The Financial Times*. https://www.ft.com/content/d71729a3-72e8-490c-bd7e-757027f9b226.

———. 2023. "Making Charts That Make an Impact." Invited talk, Data Visualization Society's Outlier Conference. https://www.youtube.com/watch?v=tIbaQUo6H9g&ab_channel=DataVisualizationSociety.

Butler, D., and S. Van Beek. 1990. "Why Not Swing? Measuring Electoral Change." *Political Science & Politics* 23 (2): 178–84. https://doi.org/10.2307/420065.

Charlotte Muth, L. 2018. "Your Friendly Guide to Colors in Data Visualisation: An Overview of Color Tools." Datawrapper. https://www.datawrapper.de/blog/colorguide.

Cleveland, W., and R. McGill. 1984. "Graphical Perception: Theory, Experimentation, and Application to the Development of Graphical Methods." *Journal of the American Statistical Association* 79 (387): 531–54. https://doi.org/10.2307/2288400.

Comber, A., C. Brunsdon, M. Charlton, G. Dong, R. Harris, B. Lu, Y. Lü, et al. 2023. "A Route Map for Successful Applications of Geographically Weighted Regression." *Geographical Analysis* 55 (1): 155–78. https://doi.org/10.1111/gean.12316.

Correll, M., and M. Gleicher. 2014. "Error Bars Considered Harmful: Exploring Alternate Encodings for Mean and Error." *IEEE Transactions on Visualization and Computer Graphics* 20 (12): 2142–51. https://doi.org/10.1109/TVCG.2014.2346298.

Donoho, D. 2017. "50 Years of Data Science." *Journal of Computational and Graphical Statistics* 26 (6): 745–66. https://doi.org/10.1080/10618600.2017.1384734.

Dykes, J., and C. Brunsdon. 2007. "Geographically Weighted Visualization: Interactive Graphics for Scale-Varying Exploratory Analysis." *IEEE Transactions on Visualization and Computer Graphics* 13 (6): 1161–68. https://doi.org/10.1109/tvcg.2007.70558.

Financial Times. 2020. "Coronavirus Trajectory Tracker Explained." https://www.ft.com/video/9a72a9d4-8db1-4615-8333-4b73ae3ddff8.

———. 2021. "'It's not a bad flu season' - Covid myths debunked with data." https://www-ft-com.proxy-ub.rug.nl/video/0cd6f9f9-664e-40f9-bad4-dde59d7c746c?playlist-name=latest&playlist-offset=223.

Franconeri, S. L., L. M. Padilla, P. Shah, J. M. Zacks, and J. Hullman. 2021. "The Science of Visual Data Communication: What Works." *Psychological Science in the Public Interest* 22 (3): 110–61. https://doi.org/10.1177/15291006211051956.

Friendly, M. 1992. "Mosaic Displays for Loglinear Models." In *ASA, Proceedings of the Statistical Graphics Section*, 61–68.

Gamio, L., and D. Keating. 2016. "How Trump Redrew the Electoral Map, from Sea to Shining Sea." The Washington Post. https://www.washingtonpost.com/graphics/politics/2016-election/election-results-from-coast-to-coast/.

Gelman, A. 2004. "Exploratory Data Analysis for Complex Models." *Journal of Computational and Graphical Statistics* 13 (4): 755–79. https://doi.org/10.1198/106186004X11435.

Gelman, A., and J. Hill. 2006. *Data Analysis Using Regression and Multilevel/Hierarchical Models*. Cambridge, UK: Cambridge University Press. https://doi.org/10.1017/CBO9780511790942.

Gelman, A., J. Hill, and A. Vehtari. 2020. *Regression and Other Stories*. Analytical Methods for Social Research. Cambridge University Press. https://doi.org/10.1017/9781139161879.

Gelman, A., J. Hill, and M. Yajima. 2012. "Why We (Usually) Don't Have to Worry about Multiple Comparisons." *Journal of Research on Educational Effectiveness* 5 (2): 189–211. https://doi.org/10.1080/19345747.2011.6182 13.

Gleicher, M., D. Albers, R. Walker, I. Jusufi, C. Hansen, and J. Roberts. 2011. "Visual Comparison for Information Visualization." *Information Visualization* 10 (4): 289–309. https://doi.org/10.1177/1473871611416549.

Gross, J. 2016. "How to Better Communicate Election Forecasts — in One Simple Chart." The Washington Post. https://www.washingtonpost.com/n ews/monkey-cage/wp/2016/11/29/how-to-better-communicate-election-forecasts-in-one-simple-chart/.

Hanretty, C. 2017. "Areal Interpolation and the UK's Referendum on EU Membership." *Journal of Elections, Public Opinion and Parties* 37 (4): 466–83. https://doi.org/10.1080/17457289.2017.1287081.

Haroz, S., R. Kosara, and S. L. Franconeri. 2016. "The Connected Scatterplot for Presenting Paired Time Series." *IEEE Transactions on Visualization and Computer Graphics* 22 (9): 2174–86. https://doi.org/10.1109/TVCG.2 015.2502587.

Harrower, M., and C. A. Brewer. 2003. "ColorBrewer.org: An Online Tool for Selecting Colour Schemes for Maps." *The Cartographic Journal* 40 (1): 27–37. https://doi.org/10.1179/000870403235002042.

Healy, K. 2019. *Data Visualization: A Practical Introduction.* Princeton, NJ: Princeton University Press. https://socviz.co.

———. 2020. *covdata: COVID-19 Case and Mortality Time Series.* http://kjhealy.github.io/covdata.

Heer, J., and M. Bostock. 2010. "Crowdsourcing Graphical Perception: Using Mechanical Turk to Assess Visualization Design." In *ACM Human Factors in Computing Systems*, 203–12. https://doi.org/10.1145/1753326.1753357.

Henry Riche, N., C. Hurter, N. Diakopoulos, and S. Carpendale, eds. 2018. *Data-Driven Storytelling.* Abingdon, UK: CRC Press. https://doi.org/10.1 201/9781315281575.

Hullman, J., and A. Gelman. 2021. "Designing for Interactive Exploratory Data Analysis Requires Theories of Graphical Inference." *Harvard Data Science Review* 3 (3).

Hullman, J., P. Resnick, and E. Adar. 2015. "Hypothetical Outcome Plots Outperform Error Bars and Violin Plots for Inferences About Reliability of Variable Ordering." *PLOS One* 10 (11). https://doi.org/10.1371/journal.pone.0142444.

Ismay, C., and A. Kim. 2020. *Statistical Inference via Data Science: A ModernDive into R and the Tidyverse.* New York, NY: CRC Press. https://doi.org/10.1201/9780367409913.

Jeppson, H., and H. Hofmann. 2023. "Generalized Mosaic Plots in the ggplot2 Framework." *The R Journal* 14 (4): 50–78. https://doi.org/10.32614/RJ-2023-013.

Kale, A., F. Nguyen, M. Kay, and J. Hullman. 2019. "Hypothetical Outcome Plots Help Untrained Observers Judge Trends in Ambiguous Data." *IEEE Transactions on Visualization and Computer Graphics* 25 (1): 892–902. https://doi.org/10.1109/TVCG.2018.2864909.

Kay, M. 2021. "Uncertainty Visualization as a Moral Imperative." Invited talk, BostonCHI meeting. https://www.youtube.com/watch?v=mfQ3QVyw4N0&ab_channel=BostonCHI.

———. 2024. "`ggdist`: Visualizations of Distributions and Uncertainty in the Grammar of Graphics." *IEEE Transactions on Visualization and Computer Graphics* 30 (1): 414–24. https://doi.org/10.1109/TVCG.2023.3327195.

Kinkeldey, C., A. MacEachren, and J. Schiewe. 2014. "How to Assess Visual Communication of Uncertainty? A Systematic Review of Geospatial Uncertainty Visualisation User Studies." *The Cartographic Journal* 51 (4): 372–86. https://doi.org/10.1179/1743277414Y.0000000099.

Kosara, R. 2023. "Lesson 4: Presentation, Uncertainty, ISOTYPE." ObservableHQ Notebook; ObservableHQ. https://observablehq.com/@observablehq/lesson-4-presentation-uncertainty-isotype?collection=@observablehq/advanced-data-vis-course.

Kuhn, M., and J. Silge. 2023. *Tidy Modelling with R*. Sebastopol, CA: O'Reilly.

Lawlor, O., and H. Robertson. 2021. "Masters series: Maarten Lambrechts' connected scatter plot." https://flourish.studio/blog/masters-connected-scatter-maarten-lambrechts/.

Lovelace, R., M. Morgan, L. Hama, M. Padgham, D. Ranzolin, and A. Sparks. 2019. "Stats 19: A Package for Working with Open Road Crash Data." *The Journal of Open Source Software* 4 (33): 1181. https://doi.org/10.21105/joss.01181.

Lovelace, R., J. Nowosad, and J. Muenchow. 2019. *Geocomputation with R*. London, UK: CRC Press.

Loy, A., H. Hofmann, and D. Cook. 2017. "Model Choice and Diagnostics for Linear Mixed-Effects Models Using Statistics on Street Corners." *Journal of Computational and Graphical Statistics* 26 (3): 478–92. https://doi.org/10.1080/10618600.2017.1330207.

McGill, Tukey, R., and W. A. Larsen. 1978. "Variations of Box Plots." *The American Statistician* 32: 12–16. https://doi.org/10.2307/2683468.

Munzner, T. 2014. *Visualization Analysis and Design*. AK Peters Visualization Series. Boca Raton, FL: CRC Press.

NHC. 2023. "National Hurricane Center and Central Pacific Hurricane Center." https://www.nhc.noaa.gov/.

Noble, S., D. McLennan, M. Noble, E. Plunkett, N. Gutacker, M. Silk, and G. Wright. 2019. "The English Indices of Deprivation 2019." Ministry of Housing, Communities & Local Government. https://www.gov.uk/government/statistics/english-indices-of-deprivation-2019.

Open Science Collaboration. 2015. "Estimating the Reproducibility of Psychological Science." *Science* 349 (6251): aac4716. https://doi.org/10.1126/science.aac4716.

Padilla, L., M. Kay, and J. Hullman. 2021. "Uncertainty Visualization." In *Wiley StatsRef: Statistics Reference Online*, edited by B. Everitt N. Balakrishnan T. Colton and J. L. Teugels. Wiley. https://doi.org/10.1002/9781118445112.stat08296.

Pebesma, E. 2018. "Simple Features for R: Standardized Support for Spatial Vector Data." *The R Journal* 10 (1): 439–46. https://doi.org/10.32614/RJ-2018-009.

Roth, R. 2021. "Cartographic Design as Visual Storytelling: Synthesis and Review of Map-Based Narratives, Genres, and Tropes." *The Cartographic Journal* 58 (1): 83–114. https://doi.org/10.1080/00087041.2019.1633103.

Scherer, C. 2023. "Designing Data Visualizations to Successfully Tell a Story." Workshop at Posit::conf(2023), Chicago, IL. https://posit-conf-2023.github.io/dataviz-storytelling/.

Silver, N. 2016. "Why FiveThirtyEight Gave Trump a Better Chance Than Almost Anyone Else." FiveThirtyEight. https://fivethirtyeight.com/features/why-fivethirtyeight-gave-trump-a-better-chance-than-almost-anyone-else.

Stevens, S. 1946. "On the Theory of Scales of Measurement." *Science* 103 (2684): 677–80. https://doi.org/10.1126/science.103.2684.677.

The New York Times. 2021. *Coronavirus (Covid-19) Data in the United States.* https://github.com/nytimes/covid-19-data.

The Turing Way Community. 2025. "The Turing Way: A Handbook for Reproducible, Ethical and Collaborative Research." Zenodo. https://doi.org/10.5281/zenodo.15213042.

Thebault, R., and A. Hauslohner. 2020. "Covid-19's Deadly New Surge Is in Rural America as States Re-Open - The Washington Post." The Washington Post. https://www.washingtonpost.com/nation/2020/05/24/coronavirus-rural-america-outbreaks/?arc404=true.

Tortosa, E. V., R. Lovelace, E. Heinen, and R. P. Mann. 2021. "Socioeconomic Inequalities in Cycling Safety: An Analysis of Cycling Injury Risk by Residential Deprivation Level in England." *Journal of Transport & Health* 23: 101291. https://doi.org/10.1016/j.jth.2021.101291.

Tufte, E. 1983. *The Visual Display of Quantitative Information.* Cheshire, CT: Graphics Press.

Tukey, J. W. 1962. "The Future of Data Analysis." *The Annals of Mathematical Statistics* 33 (1): 1–67. https://doi.org/10.1214/aoms/1177704711.

Tukey, John W. 1977. *Exploratory Data Analysis.* Reading, MA: Addison-Wesley.

Uberoi, E., C. Baker, and R. Cracknell. 2020. "General Election 2019: Full Results and Analysis." House of Commons Library. https://commonslibrary.parliament.uk/research-briefings/cbp-8749/.

Van Goethem, A., A. Reimer, B. Speckmann, and J. Wood. 2014. "Stenomaps: Shorthand for Shapes." *IEEE Transactions on Visualization and Computer Graphics* 20 (12): 2053–62. https://doi.org/10.1109/TVCG.2014.2346274.

Visvalingam, M. 1981. "The Signed Chi-Score Measure for the Classification and Mapping of Polychotomous Data." *The Cartographic Journal* 18 (1): 32–43. https://doi.org/10.1179/caj.1981.18.1.32.

White, T. 2017. "Symbolization and the Visual Variables." In *The Geographic Information Science & Technology Body of Knowledge*, edited by John P. Wilson.

Wickham, H. 2010. "A Layered Grammar of Graphics." *Journal of Computational and Graphical Statistics* 19 (1): 3–28. https://doi.org/10.1198/jcgs.2009.07098.

———. 2014. "Tidy Data." *Journal of Statistical Software* 59 (10): 1–23. https://doi.org/10.18637/jss.v059.i10.

Wickham, H., M. Çetinkaya-Rundel, and G. Grolemund. 2023. *R for Data Science: Import, Tidy, Transform, Visualize, and Model Data*. Second. Sebastopol, CA: O'Reilly Media.

Wickham, H., D. Cook, H. Hofmann, and A. Buja. 2010. "Graphical Inference for Infovis." *IEEE Transactions on Visualization and Computer Graphics* 16 (6): 973–79. https://doi.org/10.1109/TVCG.2010.161.

Wickham, H., and G. Grolemund. 2017. *R for Data Science: Import, Tidy, Transform, Visualize, and Model Data*. Sebastopol, CA: O'Reilly Media.

Wickham, H., D. Navarro, and T. Lin Pedersen. 2023. *ggplot2: Elegant Graphics for Data Analysis*. 3rd ed. New York, NY: Springer.

Wilkinson, L. 1999. *The Grammar of Graphics*. New York, NY: Springer.

Wolf, L. J., L. Anselin, D. Arribas-Bel, and L. Rivers Mobley. 2021. "On Spatial and Platial Dependence: Examining Shrinkage in Spatially Dependent Multilevel Models." *Annals of the American Association of Geographers* 111 (6): 1679–91. https://doi.org/10.1080/24694452.2020.1841602.

Wood, J., J. Dykes, and A. Slingsby. 2010. "Visualisation of Origins, Destinations and Flows with OD Maps." *The Cartographic Journal* 47 (2): 117–29. https://doi.org/10.1179/000870410X12658023467367.

Wood, J., P. Isenberg, T. Isenberg, J. Dykes, N. Boukhelifa, and A. Slingsby. 2012. "Sketchy Rendering for Information Visualization." *IEEE Transactions on Visualization and Computer Graphics* 18 (12): 2749–58. https://doi.org/10.1109/TVCG.2012.262.

Wood, J., A. Kachkaev, and J. Dykes. 2018. "Design Exposition with Literate Visualization." *IEEE Transactions on Visualization and Computer Graphics* 25 (1): 759–68. https://doi.org/10.1109/TVCG.2018.2864836.

Wood, J., A. Slingsby, and J. Dykes. 2011. "Visualizing the Dynamics of London's Bicycle-Hire Scheme." *Cartographica: The International Journal for Geographic Information and Geovisualization*, no. 4: 239–51. https://doi.org/10.1179/000870410X12658023467367.

Yang, F., M. Cau, C. Mortenson, H. Fakhari, A. D. Lokmanoglu, J. Hullman, S. Franconeri, N. Diakopoulos, E. C. Nisbet, and M. Kay. 2024. "Swaying the

Public? Impacts of Election Forecast Visualizations on Emotion, Trust, and Intention in the 2022 U.S. Midterms." *IEEE Transactions on Visualization and Computer Graphics* 30 (1): 23–33. https://doi.org/10.1109/TVCG.2023.3327356.

Yang, Y., R. Beecham, A. Heppenstall, A. Turner, and A. Comber. 2022. "Understanding the Impacts of Public Transit Disruptions on Bikeshare Schemes and Cycling Behaviours Using Spatiotemporal and Graph-Based Analysis: A Case Study of Four London Tube Strikes." *Journal of Transport Geography* 98: 103255. https://doi.org/10.1016/j.jtrangeo.2021.103255.

# Index

For Product Safety Concerns and Information please contact our EU
representative  GPSR@taylorandfrancis.com
Taylor & Francis Verlag GmbH, Kaufingerstraße 24, 80331 München, Germany

www.ingramcontent.com/pod-product-compliance
Lightning Source LLC
Chambersburg PA
CBHW070713220326
41598CB00024BA/3125